MICROBIOLOGY OF LANDFILL SITES

Second Edition

Edited by

ERIC SENIOR, Ph.D.

Waste-tech Professor of Waste Technology
International Centre for Waste Technology (Africa)
University of Natal
Pietermaritzburg, Republic of South Africa

LEWIS PUBLISHERS
Boca Raton   Ann Arbor   London   Tokyo

Library of Congress Cataloging-in-Publication Data

Microbiology of landfill sites / editor, Eric Senior. —2nd ed.
p.   cm.
Includes bibliographical references and index.
ISBN 0-87371-968-9
   1. Sanitary landfills.   2. Refuse and refuse disposal—Biodegradation.
   I. Senior, Eric, 1948–
TD795.7.M53   1995
629.4′4564—dc20   94-27855
                                                                                    CIP

© 1995 by CRC Press, Inc.
Lewis Publishers is an imprint of CRC Press

No claim to original U.S. Government works
International Standard Book Number 0-87371-968-9
Library of Congress Card Number 94-27855
Printed in the United States of America 1 2 3 4 5 6 7 8 9 0
Printed on acid-free paper

To Marijke and Ms W.

The Editor

Eric Senior, Ph.D., is Waste-tech Professor of Waste Technology in the Department of Microbiology and Plant Pathology at the University of Natal in Pietermaritzburg, South Africa. He is also Director of the International Centre for Waste Technology (Africa).

Professor Senior read Botany at the University of Liverpool, England from which he graduated in 1974. His Ph.D. degree on the "Characterisation of a Microbial Association Growing on the Herbicide Dalapon", was awarded by the University of Kent, England in 1978. After postdoctoral fellowships at the University of Groningen, The Netherlands and the University of Essex, England, Professor Senior was appointed Lecturer in Applied Microbiology at the University of Strathclyde in Scotland.

In 1990 he left to take the Foundation Chair of Microbiology at the University of Natal. Professor Senior relinquished this Chair in 1992 to take his present position.

Professor Senior is a member of the Society for General Microbiology and has served on Fermentation and Ecology Group Committees and its Scottish branch. He is also a member of the Society for Applied Bacteriology and the South African Society for Microbiology.

Internationally, Professor Senior has held Visiting Chairs and acted as advisor/consultant in a number of countries including the U.K., Malaysia, Singapore, Indonesia, Hong Kong, and India. He is also a member of the editorial board of the World Journal of Microbiology and Biotechnology.

Professor Senior is an author of more than 200 publications. His current research interests focus on the fundamental interspecies interactions underpinning environmental biotechnologies and the elucidation of these by use of microcosms and laboratory models.

Preface

Due mainly to their own efforts, many research scientists live in a generous ''comfort zone'' whereby their science is, perhaps, as much as 25 years ahead of its application. In waste technology, however, there is seldom even a positive margin and little of the biotechnology is underpinned by fundamental knowledge. This situation is increasingly exacerbated by the multi-disciplinary nature of the subject.

Edition one of this series addressed the paucity of research of the microbiology of landfill sites. For Edition two it would have been all too easy to update the first text. To do this, however, would have undermined the commitment and focus of the microbiologists who are attempting to elucidate the complex solid-state refuse fermentation. This text, therefore, not only considers the latest findings in landfill leachate treatment, co-disposal, and fundamental microbiology (by use of laboratory models), but also brings together the expertise of the immediate complementary, but often disparate, disciplines of soil science, environmental engineering, applied mathematics, and land reclamation to focus on the common goal of the scientific design and management of landfill sites.

Eric Senior
Pietermaritzburg, 1994

Contributors

Geoffrey E. Blight, Ph.D.
Department of Civil and
  Environmental Engineering
University of the Witwaterstrand
Johannesburg, South Africa

Trevor J. Britz, D.Sc.
Department of Microbiology and
  Biochemistry
University of the Orange Free
  State
Bloemfontein, South Africa

Chris A. du Plessis, B.Sc.
Department of Agronomy
University of Natal
Pietermaritzburg, South Africa

**Jeff C. Hughes, B.Sc., M.Sc.,
  Ph.D.**
Department of Agronomy
University of Natal
Pietermaritzburg, South Africa

L. Robin Jones, B.Sc., Ph.D.
Biological Laboratory
University of Kent
Canterbury, Scotland

Eric Senior, Ph.D.
International Centre for Waste
  Technology (Africa)
University of Natal
Pietermaritzburg, South Africa

Kevin J. Sinclair, B.Sc., Ph.D.
Department of Biological and
  Chemical Sciences
Bell College of Technology
Hamilton, Scotland

**Irene A. Watson-Craik, B.Sc.,
  Ph.D.**
Department of Bioscience
  and Biotechnology
University of Strathclyde
Glasgow, Scotland

Alan Young, B.A., M.A., D.Phil.
Department of Clinical Medicine
Wellcome Genetics Centre
Nuffield Hospital
Oxford, England

**Peter J. K. Zacharias, B.Sc.,
  M.Sc.**
Department of Grassland Science
University of Natal
Pietermaritzburg, South Africa

Contents

CHAPTER 1

On Isolating a Landfill from the Surrounding Water Regime

Geoffrey E. Blight

CONTENTS

I. INTRODUCTION

One of the main environmental concerns of landfilling is that polluted water may escape the confines of the landfill and enter either the surface water or the groundwater environment. This chapter will deal with the measures necessary to prevent escapes of this sort from occurring.

0-87371-968-9/95/$0.00+$.50
© 1995 by CRC Press, Inc.

The three main topics to be considered are

- The control of surface water
- Circumstances under which leachate will be generated
- The control of seepage of leachate from a landfill

II. CONTROL OF SURFACE WATER

The control of surface water is very simple to carry out effectively, but often tends to be overlooked in the siting and design of a landfill. The essential points to note in siting a landfill relevant to the control of surface water are the following:

1. No landfill should be constructed across a valley, water course, or normally dry drainage channel that may direct water onto, into, or under the landfill.
2. No landfill should be constructed so that any part of it is situated below the 50-year flood level or closer than 100 m to any stream or body of surface water.

A landfill will always lie within some or other surface water catchment. It should be isolated from the natural catchment by means of a stormwater cutoff trench, as illustrated by Figure 1. This effectively excises the landfill from its natural catchment and allows all precipitation that falls within the confines of the landfill to be retained therein. If this water has become polluted by contact with the waste, it can be used within the confines of the landfill site for irrigation of grass or other vegetation or to keep down dust on roads and working areas. Alternatively, the polluted water can be treated before discharge from the site into a natural watercourse.

The stormwater cutoff drain is usually sized so that it can divert the runoff from a specific design rainstorm for the area in which the landfill is located, while maintaining a specified freeboard. For example, the design storm may be the 50-year storm of 24-h duration, for which the runoff must be diverted while a minimum freeboard of 0.5 m is maintained.

Figure 1 also indicates a polluted water runoff collection trench which should be designed to meet similar size and flow criteria.

III. RUNOFF AND INFILTRATION INTO COVERED LANDFILL SURFACES

One of the most important aspects concerning water control on landfills is the proportion of rainfall falling on a soil-covered landfill surface that

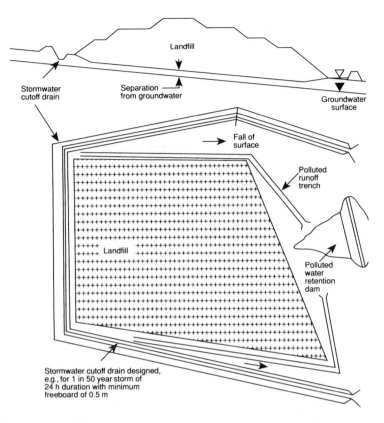

Figure 1 Principle of separating a landfill from external surface water and groundwater.

runs off and enters the surface water regime and the corresponding proportion that infiltrates the cover layers. Of the water that infiltrates, some will in general be reevaporated or evapotranspired and some will continue to move into the refuse body, possibly to emerge at the base of the landfill as leachate. Infiltration is related to precipitation, runoff, and interception by the equation

$$\text{Infiltration} = \text{precipitation} - (\text{runoff} + \text{interception}) \tag{1}$$

Interception comprises water that is intercepted above the surface of the cover layer by vegetation and either evaporates or reaches the surface at a later time. Equation 1 indicates how it is possible to measure infiltration: by deducting measured runoff from measured precipitation (making allowance for interception) or by directly measuring the gain in moisture of the cover layers following precipitation.

**Table 1   Comparison of Average Steady-State
Saturated Infiltration Rates for
De-ionized Water and Domestic Supply**

Test no.	De-ionized water (mm/h)	Domestic supply (mm/h)
1	25	30
2	40	42
3	1	2.5
4	No test[a]	75

Note: Mean coefficient of permeability measured in laboratory: 1.4 mm/h

[a] Infiltration rate too rapid for available de-ionized water supply.

It can be argued that the steady-state saturated infiltration rate, as measured by means of a double-ring infiltrometer, would represent the maximum rate at which precipitating rain could be absorbed by a saturated soil surface. Rainfall with an intensity of less than the saturated infiltration rate would infiltrate entirely, while, at intensities greater than the saturated infiltration rate, the surplus would run off. If the surface is vegetated, rain may be intercepted by the vegetation and either evaporate without reaching the soil surface or reach the soil at a rate slower than the rainfall rate. If the soil surface is not saturated, the rate of infiltration can exceed the steady-state saturated infiltration rate until the surface layer becomes saturated. This is because, initially, no runoff can take place until all the slight depressions in the soil surface are full and start to overflow. Also, because of suction or negative pore water pressures within an unsaturated soil, the infiltration flow gradient for an unsaturated soil will exceed the infiltration flow gradient for a saturated soil.

However, the real problems in estimating infiltration are much more complex than the above discussion would indicate. Recent research[1] has shown the following.

The saturated steady-state infiltration rate is heavily dependent on the state of the soil surface, mainly on whether or not the surface is cracked when precipitation falls on it. For example, average steady-state saturated infiltration rates recorded for four sets of double-ring infiltrometer tests carried out on the final cover layer of a landfill (the Linbro Park landfill in Johannesburg, South Africa) are shown in Table 1. Of the four tests, the positions for tests 1 and 2 showed slight surface cracking, that for test 3 was uncracked, while the position for test 4 was crossed by a wide and deep crack.

Ideally, the water used for infiltration measurements should be local rainwater or water with a chemistry similar to that of local rain. In this case, such water was not available, and it was necessary to use the local domestic water supply to carry out the tests. Table 2 compares the properties of the

Table 2   Comparison of Chemistry of Domestic Water
Supply and of Rain at Experimental Site

Parameter	Rain	Local supply
Total hardness	5 mg/l	100 mg/l
Calcium hardness	3 mg/l	67 mg/l
pH	4.2	7.5
Conductivity	2 mS/m	33 ms/m
Chloride	Trace	18 mg/l
Sulphate	3 mg/l	40 mg/l
Calcium	4 mg/l	27 mg/l
Sodium	Trace	21 mg/l

local domestic water supply with that of local rain. It will be noted that the two waters differ considerably in properties.

Reference to work by Agassi et al.[2] indicated that the infiltration behavior of the soil at the experimental site might be sensitive to the chemistry of the infiltrating water. This could cause the hydraulic conductivity to increase. In order to check this, double-ring infiltrometer tests were undertaken with both de-ionized water (to simulate rainwater) and with the domestic water supply.

Table 1 shows that, while there was a tendency for the infiltration of domestic supply water to be more rapid than that of distilled water, the difference was not large. The saturated steady-state infiltration rates for the positions of tests 1, 2, and 4 which contained surface cracks were considerably higher than the value of the coefficient of permeability measured in the laboratory. However, for the uncracked position for test 3, the infiltration rate was almost the same as the permeability coefficient measured on intact specimens in the laboratory.

According to the above argument, the excess of the actual rainfall intensity over the steady-state saturated infiltration rate into the landfill cover would run off. However, the actual intensity for a rainfall event is not usually known and, hence, the runoff cannot easily be evaluated.

Most studies of rainfall patterns have been carried out for the purposes of storm hydrology (e.g., see Reference 3). In this case the heaviest rainfall and the greatest runoff over a period of years is of interest (e.g., in the case of designing stormwater diversion drains). When considering infiltration into landfill cover layers, however, it is the everyday rainfall events that are of most interest as they will contribute the greatest quantity of infiltration. Likewise, most methods for estimating runoff are based on exceptional or storm rainfall events, whereas the interest for landfills centers about common rainfall events.

Figure 2 shows an analysis of a 30-year record of daily rainfall for a weather station in Johannesburg. This shows that 60% of daily rainfall depths are less than 5 mm, 75% are less than 10 mm, and 90% are less than 20 mm. These are the common events that can be expected to cause most infiltration into the landfill.

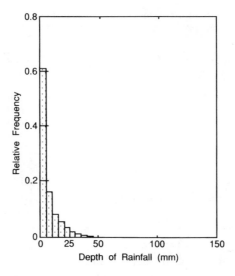

Figure 2   Frequency-depth distribution for daily rainfall observed near landfill site.

Because of the lack of information on runoff and infiltration, simulations have been made of a series of rainfall events of 5-, 10-, and 20-mm depths. As no information was available concerning the distribution of rainfall intensity within a particular event, it was decided to adopt approximations to the distributions proposed by the U.S. Soil Conservation Service (SCS), as modified for local conditions by Schmidt and Schulze.[4] The approximated SCS rainfall intensity distributions for the design storms of 5, 10, and 20 mm are shown in Figure 3. Figure 3 also shows the actual SCS distribution for the 20-mm event.

The SCS distributions are based on storms having 10- to 20-year recurrence intervals. With more frequent storms, the peak rainfall intensity is known usually to occur closer to the start of the storm, rather than in the middle as assumed by the SCS. The SCS distributions and especially the truncated approximations used in this work (Figure 3) would tend to result in an overestimation of actual infiltration.

A cheap, easily portable sprinkler system was designed to provide a uniform distribution of irrigation at the design rainfall intensities required (3, 5, 7, 10, and 20 mm/h). The design consisted of four commercially available irrigation sprinklers located at the corners of a square or rectangle so that the sprinklers were symmetrically disposed about the 9 m × 9 m square test plot. Depending on the required intensity, the sprinklers were spaced at distances varying from 18 m × 18 m to 9 m × 9 m. In each case, the test plot lay in the area of overlap of the four sprinklers, which were mounted with their spray nozzles at 1 m above the ground surface. The water supply was contained in a collapsible portable plastic reservoir of

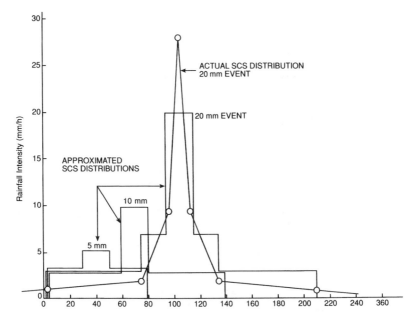

Figure 3   Rainfall intensity for sprinkler infiltrometer rainfall simulations. Actual SCS distribution shown for 20-mm event.

6 m³ capacity. The system was fed by a portable centrifugal pump capable of a maximum head of 600 kPa and a maximum flow rate of 24 m³/h.

Five rain gauges were positioned randomly over the surface of each test plot to measure the actual application of water. The tests were performed during clear, warm, dry weather, and the rain gauges were found to collect only 70 to 75% of the design storm depth. This resulted from evaporation of water into the air in the time from when it left the nozzle to when it reached the ground.

Each test plot was surrounded by a system of impervious plastic gutters positioned so that they diverted any water that tended to run onto the test plot from outside and intercepted all water that ran off the test plot. Runoff from the plot collected in the gutters and was channeled to a collecting box at the lowest corner of the plot. From here it was piped to a measuring drum.

The comparison shown in Table 3 shows that the SCS method (based on empirical curves) gives predictions of runoff that are surprisingly accurate since the method was derived for considerably more severe storm events than those simulated in this study. It is, therefore, recommended that the SCS curves be used to assess runoff from covered landfill surfaces. Once the partition between runoff and infiltration has been made, design runoff values can be used to size the polluted runoff trenches indicated in

Table 3   Summary of Results of Runoff Tests Using Sprinkler Infiltrometer

Plot no.	Antecedent moisture content (% dry mass)	Vegetation % and cracking of surface	Estimated effective simulated rainfall depth (mm)	% Runoff[a]	
				Measured (%)	SCS prediction (%)
One	2	75%	4.1	0	0
1.7%	2		8.3	0.4	0
slope				0.3	
	2	No cracks	16.1	2.5	0
				1.8	
	14		4.1	1.7	5–7
				1.2	
Two	2	50%	4.1	0	0.1
1.6%	2		8.3	0.3–0.2	0.2–0.3
slope	2	Cracks	16	4.7–3.8	0.4–0.5
Three	1	100%	4.1	0	0.1
20%	1	burnt	8.3	0.4–0.3	0.2–0.3
slope	2	No cracks	18	0.8–0.7	0.4–0.5

[a] A range of runoff percentages is given. The first figure is the runoff as a percentage of the rainfall depth measured in the rain gauges. The second figure is the runoff as a percentage of the design rainfall depth. The true figure lies between these two.

Figure 1. The water that infiltrates the landfill cover layer will either evaporate or be evapotranspired or will seep into the body of the refuse.

The most important conclusion from Table 3, however, is that under rainfall of low intensity the percentage runoff is very low and could be ignored in design. As Figure 2 shows that most of the annual rainfall at the test site occurs as rain of low intensity, the annual runoff will also be very low.

Without investigating this statement, it is very likely that this conclusion applies in most climates and could be adopted as a general rule for design.

IV. THE WATER BALANCE FOR A LANDFILL

A great deal of concern exists that existing and future sanitary landfills are causing, or have the potential to cause, unacceptable groundwater pollution. Such pollution is most costly and difficult to remediate once it has occurred. If nothing is done to ameliorate the situation, the pollution may persist in the groundwater for decades[5] even though the source of the pollution has been removed.

A preliminary study[5,6] produced strong evidence that, if climatic conditions are such that a perpetual water deficit exists at the site of a landfill, no or very little leachate will be formed or exit from the base of the landfill. Hence, if there is an adequate separation between the lowest level of refuse

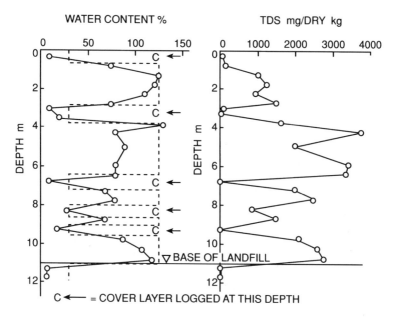

Figure 4   Water content and TDS profiles for landfill at end of dry season.

and the highest level of the regional phreatic surface, no groundwater pollution will occur. By extension, surface water replenished by the groundwater will also remain unpolluted by leachate from the landfill.

Figure 4 shows profiles of water content and total dissolved solids (TDS) measured in a landfill in Johannesburg at the end of the dry season. Johannesburg has a perennial climatic water deficit. It will be noted that not only is the soil underlying the landfill relatively dry but the TDS in this soil is close to zero. This shows that pollution is not escaping from this landfill, even though the landfill was 10 years old when the data of Figure 4 were obtained.

There is considerable support for this view in the literature. For example, Keenan[7] gives figures indicating that landfills receiving more than 750 mm of precipitation per annum will eventually produce leachate, while those in arid regions receiving less than an annual 325 mm are likely never to exude pollution. Saxton[8] states that, for climates where annual precipitation is less than 400 mm, virtually all precipitation is evapotranspired. Earlier, Fenn et al.,[9] Burns and Karpinski,[10] and Holmes[11] all agreed that, if a net annual water deficit exists at the site of a landfill, little leachate will exit from its base.

It must also be recognized that good engineering and management of a landfill can be used to maintain a perennial water deficit within the fill even though there may actually be an excess of precipitation over potential evaporation. This can be done by maximizing runoff and minimizing infiltration

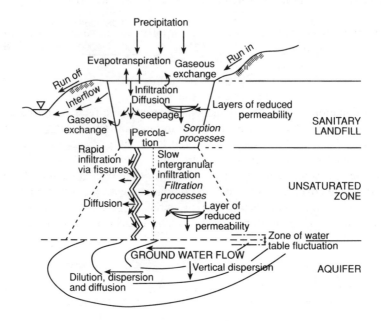

Figure 5 Details of the components of a water balance in a sanitary landfill.

into the refuse. A suitably sloping surface and the installation of a carefully designed impervious cover layer (see later) can achieve this (e.g., see Reference 12).

Whether or not a landfill in a particular climate will produce leachate can be investigated by means of the water balance which applies the principle of conservation of mass to water moving into, out of, and through a landfill. The water balance for a landfill can be stated as follows:

$$\text{water input to landfill} = \text{water output} + \text{water retained in refuse body} \quad (2)$$

Figure 5 (after Reference 13) illustrates the components of the water balance for a landfill in greater detail.

In Equation 2, each term represents a rate of accumulation or loss. Water input includes precipitation (P) and the water content of the incoming waste (U_w). U_w, however, only makes a once-off contribution to the annual water balance of a given mass of landfill. Water output includes evapotranspiration (ET), water lost in leachate escaping or removed from the landfill (L), and runoff (R). Finally, there is water absorbed and retained by the waste (ΔU_w) and the soil cover (ΔU_s). For an annual water balance:

$$P + U_w = ET + L + R + \Delta U_w + \Delta U_s \quad (3)$$

The annual water balance equation, as applied to an established landfill, can be simplified to:

$$P = ET + L + R + \Delta U_w + \Delta U_s \tag{4}$$

In Equation 4, the only components that can be directly controlled by the engineer are the runoff R, and, by limiting infiltration, the terms ΔU_w and ΔU_s. Also, we have seen earlier that R is negligible for most rainfall events and could be ignored with little error. Reference to Figure 4 will show that the water content of the cover layer of the landfill and of the soil intermediate cover layers is small in comparison with the water content of the refuse and could probably also be ignored. With these two further simplifications, Equation 4, rearranged to give the leachate production, becomes:

$$L = P - ET - \Delta U_w \tag{5}$$

Obviously, the smaller the precipitation (P) and the larger the evapotranspiration (ET) the less potential for the generation of leachate (L). These terms are particularly favorable in water deficient areas.

U_w is known as the field capacity or moisture retention capacity of the refuse. As decomposition and compaction of refuse occurs in a landfill, the field capacity will progressively decrease from its initial value at the time of deposition. The literature records values for the field capacity of refuse that vary from 80% for fresh refuse[14] to between 63 and 74% for refuse more than 4 years old.[11] These figures obviously depend both on the composition of the refuse and the method of determining the dry mass. The results of two independent studies of the field capacity of refuse are summarized in Figure 6.[15,16] The refuse contained (on a dry weight basis) 54% organics, 23% paper, 9% glass, 8% plastic, and 6% metal. Figure 6 shows an even wider variation of field capacity than those reported by Campbell[14] and Holmes.[11]

The theoretical concept that refuse will continue to absorb moisture until the field capacity is reached, and will thereafter release moisture at the same rate as it receives it, is obviously an oversimplification. It can be deduced from Figure 6 that the field capacity of refuse can be reached because of the accumulation of moisture, because the field capacity is changing as the age and state of compaction and decomposition of the refuse increase, or by a combination of the two processes. Landfills are heterogeneous in composition and may start to release leachate long before their overall field capacity has been reached because certain interconnected zones within them have a lower field capacity than others.

In a water-deficient area, on the other hand, the field capacity may never be reached, and the result of applying Equation 5 will be a perennially negative value for L. Figure 7 shows the results of sampling an 8-year-old landfill situated in Johannesburg, which is in a perennial water-deficient

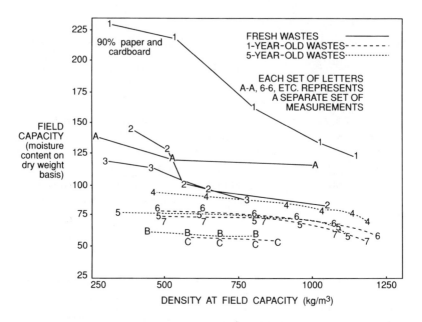

Figure 6 Measurements of the field capacity of refuse.

area.[17] The field capacity of the soil cover layers was about 30% by dry mass, while the field capacity of the refuse averaged 130%, with individual measured values going as high as 250%. As the diagram shows, at the end of the wet season, all of the refuse was well dry of its field capacity, while the cover layers had probably reached their field capacity. It will also be noted that the landfill loses moisture throughout its depth during the dry season, and regains moisture during the wet season.

Once a landfill has been established for a number of years and the refuse has approached an equilibrium moisture content, the term $\Delta \bar{U}_w$ could be omitted from Equation 5, with little effect, giving

$$L = P - ET \tag{5a}$$

which now contains only the climatic variables P and ET. Equation 5a has been called a "climatic water balance".

ET, the evapotranspiration loss from the surface of a landfill, can be evaluated by means of a variety of evapotranspiration equations, none of which appear to be completely accurate, and all of which differ in their predictions. A comparison of these equations carried out by Hojem[18] showed that the approximations

$$ET = 0.7 \times A\text{-pan evaporation} \tag{6a}$$

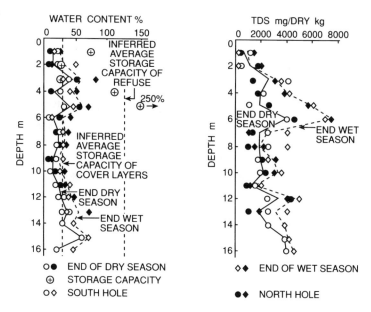

Figure 7   Water content and TDS profiles for landfill. End of wet season April 1988. End of dry season October 1988.

or

$$ET = 0.9 \times \text{S-pan evaporation} \tag{6b}$$

give estimates of ET that are as close as those of most of the better-known evaporation equations.

The climatic water balance can be used as a means of deciding whether or not a landfill will generate significant quantities of leachate and, there-fore, whether or not a leachate collection system and underliner should be provided to isolate the landfill from its surroundings. To allow for seasonal influences and variable weather patterns, L is calculated for the wet season of the wettest year on record. (The wet season would usually be taken as the wettest 6-month period in a year, based on long-term averages.) If the value of L is positive the indication is that the landfill will generate leachate in a wet year. Conversely, if L is negative the indication is that the landfill will not generate leachate even in a wet year.

As the rainfall and evaporation in any one year do not necessarily cor-relate, L is recalculated for successively drier years to establish if

1. L is positive in *less* than 1 year in 5 for which data are available.
2. L is positive in *more* than 1 year in 5. If (1) applies, a leachate collection system and underliner can safely be omitted from the landfill. If (2) ap-plies, regular generation of leachate can be expected, and a leachate col-lection system and underliner would need to be provided.

It is accepted that, even if (1) applies, a landfill may generate leachate following an exceptionally wet period or an exceptionally wet season. However, this will be a sporadic occurrence, making minimal impact on the environment and will not justify the cost of installing an underliner and leachate collection system.

V. THE DESIGN OF LANDFILL LINERS—LEACHATE COLLECTION SYSTEMS

If, during the design of a sanitary landfill, it is established by considering the water balances that leachate will be generated, it will be necessary to provide the landfill with a rationally designed underliner. It will also be necessary to contour and shape the base of the landfill so that percolating leachate is directed toward collection sumps from which it can be pumped and either recirculated or treated and discharged. Figure 8 illustrates a scheme for a leachate collection system beneath a landfill.

The collection pipes shown in Figure 8 are usually laid in a leachate collection layer consisting of pervious material (see also Figures 9 and 10). Experience in Europe[19] has shown that the pores of the leachate collection layer will block with insoluble deposits unless the collection layer is made with very coarse, single-sized gravel. Currently, single-sized gravel with a size of 38 to 50 mm is being specified for leachate collection layers.

The leachate collection pipes have also been known to block with deposited material. For this reason, the collection pipes are laid in such a way that they can be cleared periodically by rodding from a manhole or manholes installed at the end or ends of each pipe. The base of the landfill between collection pipes is graded at between 2 and 5% to allow the leachate to flow toward the pipes and to ensure that the depth of leachate over the liner never exceeds more than a few centimeters.

VI. TYPES OF LANDFILL LINER

Liners currently in use beneath landfills are of two main types:

- Compacted earth liners (with the earth possibly modified by the addition of bentonite).
- Composite compacted earth/flexible membrane liners.

Figure 9 shows two typical liner designs of the above types. Figure 9a would be suitable for a site underlain by clayey soil; the impervious element of the liner consists of this natural material, removed and re-compacted. Figure 9b would be suitable for a site underlain by more pervious sandy

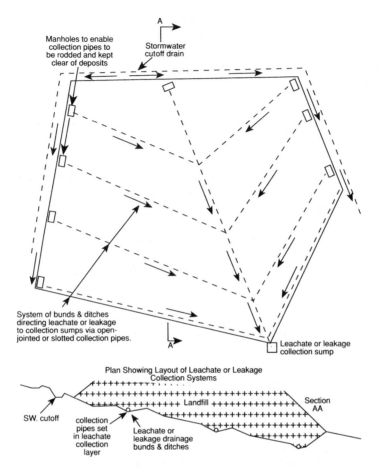

Figure 8   Principle of leachate collection system.

soils, using a flexible membrane liner laid on a prepared, compacted base. Both liners would be suitable for use with landfills storing domestic or municipal refuse.

Figure 10 shows the very much more elaborate type of liner that would be suitable for a hazardous waste landfill. An essential feature of this design is that there are three impervious elements: the upper liner of compacted earth, a flexible membrane liner, and a lower compacted earth liner. The flexible membrane liner forms a composite liner with the lower compacted earth liner.

A leakage collection layer is provided between the upper and lower liners. Its function is to collect any leakage or seepage that has passed through the upper liner and to direct it toward strategically placed leakage collection sumps. Not only does the leakage collection layer allow the

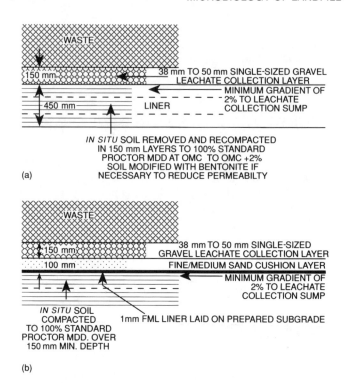

(a)

(b)

Figure 9 Examples of underliners for landfills storing domestic refuse. FML = flex-ible membrane liner, MDD = maximum dry density, and OMC = optimum moisture content.

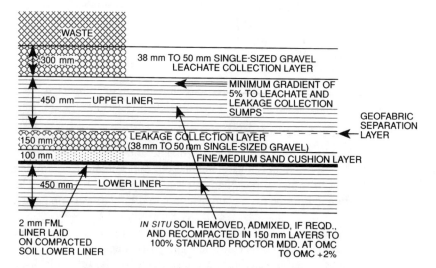

Figure 10 Liner suitable for a hazardous waste landfill. FML = flexible membrane liner, MDD = maximum dry density, and OMC = optimum moisture content.

amount of leakage to be measured, it also allows the leakage to be collected and removed from the system, so that it is not available to seep through the lower liner.

VII. POSSIBILITY AND CONSEQUENCES OF LINER FAILURE

If a compacted earth liner fails as a result of mechanical damage, desiccation cracking, or the accidental inclusion of one or several zones of pervious material, the total flow rate through the liner is unlikely to increase very much, as the damage will be localized, and the flow will be limited by the local rate of generation of leachate.

In the case of a liner retaining leachate from hazardous waste, there is the possibility that the liner may fail as a result of chemical attack by the leachate on the soil. Research has shown that alcohols, phenols, ketones, ethers, and organic bases can have pronounced deleterious effects on the permeability of clays when these organic substances are highly concentrated. Chemical attack only becomes significant at concentrations above 75% of the pure substance. In a landfill where hazardous liquids, possibly containing these organic substances in small quantities, are co-disposed with domestic refuse, it is most unlikely that highly concentrated organic substances of any description will come in contact with the earth liner. The possibility will be made even more remote because any appreciable quantities of volatile liquids of this type are normally only deposited in a landfill after absorbing them with a material like fly-ash. No free organic liquids are usually disposed in a landfill. Hence the possibility of chemical attack on a compacted earth liner is remote.

In the long term, the permeability of a compacted earth liner and, therefore, the outflow from the landfill should decrease because

- As overburden builds up on the liner, the effective stress in the liner will increase, thus decreasing the permeability.
- When the landfill is decommissioned, any addition of liquids or moisture in the waste will cease, and the generation of leachate will decline.
- The covering of the landfill with its final cap will further decrease the leachate generation.

Most sets of regulations for landfill design and construction specify a maximum seepage rate or maximum permeability for a compacted earth liner. At the design stage, this must be estimated from the results of laboratory permeability tests on the soil it is proposed to use for the liner. The methods used are described below.

VIII. CALCULATION OF THE RATE OF SEEPAGE
THROUGH A CLAY LINER

If the landfill is properly designed and operated, the depth of leachate over the liner, as controlled by the leachate collection system, should never exceed a few centimeters. In its simplest form, therefore, a clay liner must be designed for the long-term situation illustrated in Figure 11. In this figure, d_L will usually be very much less than d_C and d_F.

For the purposes of this calculation it will be assumed that one is dealing with a saturated system in which the coefficients of permeability (or hydraulic conductivity) of each layer are constant, and that Darcy's law applies with a constant coefficient of permeability: under steady-state conditions, the seepage flow through the liner and the foundation soil will be the same. By applying Darcy's law,

$$q = ki \tag{7}$$

(where k is the coefficient of permeability and i the seepage gradient).

Referring to Figure 11, and imposing the condition for continuity of flow, the outflow per unit plan area through the liner can be shown to be

$$q = \frac{k_c k_F (d_L + d_C + d_F)}{k_c d_F + k_F d_c} \tag{8}$$

If d_L is small in comparison with d_C and d_F, Equation 8 can be approximated by

$$q = \frac{k_C k_F (d_F + d_C)}{k_C d_F + k_F d_C} \tag{8a}$$

Alternatively, the liner permeability k_C is related to q by

$$k_C = \frac{q d_C}{(d_L + d_C + d_F) - \dfrac{q d_F}{k_F}} \tag{8b}$$

which can be approximated by

$$k_C = \frac{q d_C}{(d_C + d_F) - \dfrac{q d_F}{k_F}} \tag{8c}$$

if d_L is small in comparison with d_C and d_F.

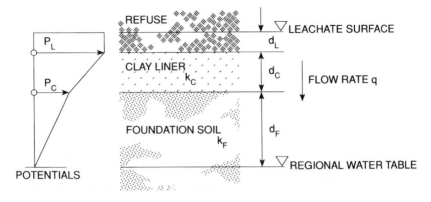

Figure 11 Basis for estimating rate of seepage through a single clay liner.

The pore water pressure at the underside of the liner is given by

$$u = \frac{(k_C - k_F)\, d_F d_C + k_C d_F d_L}{k_C d_F + k_F d_C} \tag{9}$$

or approximately by

$$u = (k_C - k_F) \left(\frac{d_F d_C}{k_C d_F + k_F d_C} \right) \tag{9a}$$

It follows from Equation 9 (and is shown particularly well by Equation 9a) that, if the permeability of the foundation soil exceeds that of the liner, the pore pressure under the liner will be negative. As the whole object of having a liner is to reduce outflow from the landfill, k_C will in practice invariably be considerably less than k_F. The negative pore pressure may result in the foundation soil becoming unsaturated immediately below the liner. Because the coefficient of permeability of a soil decreases when it becomes unsaturated, this will further decrease q.

Figure 12 shows typical calculated pore pressure profiles through a clay liner and the underlying foundation soil, as well as the modification to the pore pressure profile caused by unsaturation. The average seepage gradient over a given depth of soil is given by the change in pore pressure over that depth, plus the changes in elevation, divided by the change in elevation; for example, for the hydrostatic pore pressure line, the seepage gradient over depth d is

$$i = \frac{d - d}{-d} = 0$$

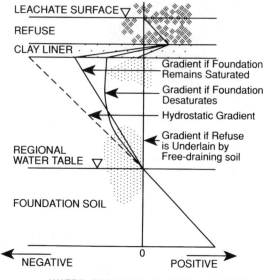

Figure 12   Typical pore pressure profiles through a clay liner.

Hence there is no flow under a hydrostatic condition. For a vertical line in Figure 12

$$i = \frac{0 - d}{-d} = 1$$

i.e., the gravity gradient of unity applies.

The average seepage gradient between the top of the clay liner and the regional water table must always be unity, but the gradient over a lesser depth (such as the liner thickness) may be much greater than unity. If k_C is very much less than k_F, and h is the change in pore pressure over depth d_C,

$$i = \frac{-h - d_C}{-d_C} = 1 + \frac{h}{d_C} \tag{10}$$

This can be very much larger than 1.

Pore pressure profiles such as those shown in Figure 12 have been measured in the laboratory; for example, Day and Luthin[20] measured negative water pressures under a less pervious layer in a soil column that were similar to those shown in Figure 12. More recently, Krapac et al.[21] measured seepage gradients in an experimental lysimeter that were as high as 1.75. Hence, there is no doubt that negative pore pressures and seepage gradients higher than unity can exist in a clay liner.

For example,

$$\text{If } k_F = 100 \ k_C \text{ and } d_F = 100 \ d_C$$

then from Equation 9a

$$u = -49.5 \ d_C$$

and from Equation 10

$$i = 50.5$$

IX. MEASUREMENT OF PERMEABILITY *IN SITU* AND IN THE LABORATORY

Although Darcy's law is usually applied to calculating the outflow through a clay liner, the literature shows that there is considerable uncertainty concerning how best to measure the Darcy coefficient of permeability (or hydraulic conductivity) k.

Day and Daniel[22] conducted comparative field and laboratory measurements of permeability coefficients on two clays. Test ponds were constructed in the field and samples were later retrieved from the test liners for laboratory measurements. Measurements of seepage rate were made for the pond as a whole and by means of single- and double-ring infiltrometers. Tests using both rigid- and flexible-walled permeameters were made on block and tube samples of the clay compacted *in situ* and also on samples compacted in the laboratory. Effective confining stresses in the laboratory were about 100 kPa and seepage gradients ranged from 20 to 200. Day and Daniel[22] found that values of permeability coefficients deduced from seepage losses from the ponds were 900 to 2000 times larger than permeability coefficients measured in the laboratory, but only 1.2 to 1.9 times larger than field infiltrometer measurements.

Chen and Yamamoto[23] also carried out a comparison of field and laboratory permeability measurements, using infiltrometers and porous probes *in situ* and flexible-walled permeameters in the laboratory. For the laboratory tests, effective stresses were about 200 kPa and the seepage gradient was 180. They found that field permeability coefficients were 10 times larger than laboratory values.

Elsbury et al.[24] made a comparison of field and laboratory permeability measurements on a highly plastic clay. They found that double-ring infiltrometer tests gave slightly lower permeability coefficients than did seepage rates from a test pond. However, compaction in the field with a vibratory roller resulted in a liner with a permeability coefficient ten times larger than one compacted using the same roller without vibration. Permeability coefficients measured in the laboratory used seepage gradients of 20 to 100 and effective stresses of 15 to 70 kPa. Permeability coefficients measured in the field proved to be between 10,000 and 100,000 times greater than values measured in the laboratory.

In contrast to the above conclusions, Pregl[25] has stated that a permeability coefficient measured in the laboratory serves as an index of material quality but is not directly related to the permeability coefficient of a lining in the field. The permeability coefficient in the field will (according to Pregl) always be less than that measured in the laboratory, because the seepage gradient used in laboratory tests is usually of the order of 30 or more whereas that in the field approximates to unity.

It is apparent from these studies that there are several possible reasons why a permeability coefficient measured in the field may differ from one measured in the laboratory and why a laboratory measurement on a large specimen may differ from a similar measurement on a small specimen:

- A large area exposed to seepage is more likely to contain defects in the form of fissures and more permeable zones than is a small area.
- If the Darcy coefficient of permeability is not constant, the use of different seepage gradients in the field and laboratory will result in different field and laboratory values.
- A similar remark applies to effective stresses. A specimen subjected to a high effective stress in a laboratory test can be expected to show a lower permeability than a similar one with a low effective stress in the field.
- The interpretation of the field permeability test may be faulty. It will be noted from Equation 8b that the permeability of the liner, k_C, cannot be evaluated without a knowledge of d_F and k_F. The literature shows various other interpretations for k_C. The most common is to assume that the seepage gradient $i = 1$, in which case $k_C = q$. Apart from the example calculation above, this is clearly not always logical because by the same argument, k_F in Figure 11 would also equal q. Daniel[26] uses the expression

$$k_C = \frac{q \, d_C}{d_L + d_C} \tag{11}$$

which is also not correct (see Equation 8c).

To get some notion of the errors involved with Daniel's expression, suppose as before that

$$d_F = 100d_C, \ d_L = d_c/10, \text{ and } k_F = 100k_C$$

If $i = 1 \ k_C = q$

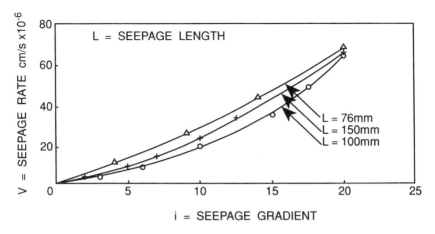

Figure 13   Typical relationships between v and i for seepage through soil.

According to Daniel's expression (Equation 11)

$$k_C = 0.91q$$

According to Equation 8c

$$k_C = 0.02q$$

Hence, conventional assumptions can result in considerable errors in interpreting the results of pond seepage tests.

One of the probable reasons, therefore, for the wide discrepancy between Daniel's field and laboratory measurements is that his field measurements may have been misinterpreted.

X. THE DARCY COEFFICIENT OF PERMEABILITY

According to the classical form of the Darcy equation (Equation 6), one is led to believe that k is a constant for all i. Pregl (1987)[25] has pointed out that this is not so for compacted clays, but rather that k increases with increasing i. The set of measurements shown in Figure 13 confirm this observation. The soil was a clayey sand residual from weathered granite and the measurements show that the seepage flow rate increases at a greater rate than the hydraulic gradient. For this set of data, the value of k at an hydraulic gradient of 1 was 1×10^{-4} cm/s while at an hydraulic gradient of 20, k was 3×10^{-4} cm/s.

Figure 13 also shows that reducing the seepage path length from 150 mm to 75 mm had relatively little effect on the seepage flow rate at a particular seepage gradient. In other words, k was not particularly sensitive to specimen size, in this case.

These results illustrate the importance of matching the seepage gradient used in laboratory tests to the gradients expected to occur in the field.

XI. POND SEEPAGE TESTS

As shown above, a field ponding test cannot be correctly interpreted in terms of the coefficient of permeability of a clay liner unless:

- The coefficients of permeability of the foundation layers have been correctly measured, or
- The pore water pressure at the underside of the clay liner is measured, or
- The depth of the foundation layer (as shown in Figures 11 and 12) is zero.

If these conditions are satisfied, far better agreement between field and laboratory tests can be obtained, as the following case study will show.

A site was selected such that d_F was effectively zero (i.e., the permeability was uniform with depth) and the position of the regional water table was known. Four 3-m square ponds were constructed on a profile of clayey sand residual from weathered granite. The water table lay at a depth of 2 m below the bottom of the ponds. The ponds were surrounded by a perimeter moat to eliminate lateral flow from the pond edges as far as possible. After shaping, the bottoms of the ponds were compacted to achieve a density similar to that of the underlying undisturbed profile. One of the four ponds was lined with a geomembrane so that it could be used to measure evaporation losses and all four ponds were filled with a coarse clinker ash to provide some overburden pressure on the pond bottoms and also to eliminate wave formation in the water and to reduce evaporation.

The rates of seepage were measured by observing the water levels in four perforated observation pipes that were installed in each pond.

Referring to Figure 11 and Equation 8b: because k_F was close in value to k_C, or alternatively, because d_F was zero, it could truly be said in this case that the average seepage gradient between the pond and the regional water table was uniform, equal to unity, and, therefore, that $k_C = q$.

Figure 14 shows a set of measured k_C values for one of the ponds. It will be seen that there is a considerable scatter in the measurements. The scatter resulted from the difficulty of measuring small changes of water level accurately and difficulty in compensating the measured seepages for evaporation losses (which were of the same order as seepage losses and also depend on measuring small changes of water level). Temperature changes also affected the accuracy of the measurements by causing the plastic measuring pipes to change in length. Thus, an increase in temperature caused an apparent increase in seepage rate, and vice versa.

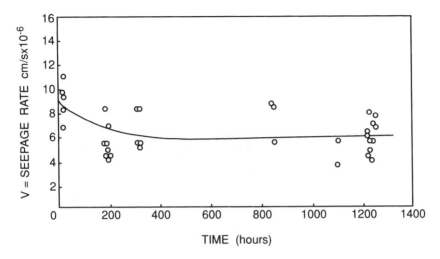

Figure 14   Measurements of seepage rate for a typical long-term pond seepage test.

Double-ring infiltrometer tests carried out at an adjacent site gave very similar values of k_C to those inferred from the ponding tests.

XII. COMPARISON OF POND AND LABORATORY PERMEABILITY TESTS

In the classic design situation the clay liner to a landfill must be designed on the basis of the results of laboratory permeability tests and be confirmed later by means of field measurements. For this reason, a comparison of pond seepage test results and laboratory permeability measurements is of particular interest. Also, as seen earlier, the laboratory tests should be carried out at a seepage gradient that is comparable with the gradient expected in the field situation, if they are to be meaningful.

Table 4 compares the permeability values measured in the field and the laboratory.

Hence this set of measurements shows that it is possible, with correct interpretation and correct testing, to estimate *in situ* permeabilities reasonably closely from the results of laboratory tests.

However, it must be noted that the permeability of the soil used in these tests is higher than what would usually be used for a landfill liner. It is also

Table 4   Comparison of Permeabilities Measured in the Field and the Laboratory (cm/s $\times$ 10^{-6})

Ponding tests	Double-ring infiltrometer	Laboratory (triaxial)
6–8	1–8	4–9

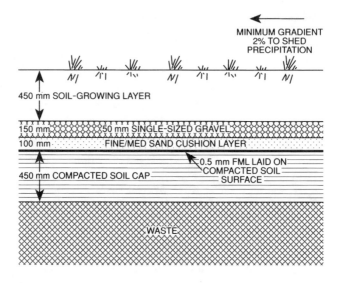

Notes:
1. Soil cap is compacted to 100% standard Proctor maximum dry density
2. Permeability of compacted soil must not exceed 0.1m/y as measured by double ring infiltrometer tests.

Figure 15   Final cap/cover layer for hazardous waste landfill.

noted from the literature that discrepancies between field and laboratory permeabilities appear to increase as the soil becomes less permeable.

XIII. THE DESIGN OF FINAL COVER OR CAPPING LAYERS TO LANDFILLS

Once a landfill has been brought to its final height and surface contours, a final cover layer, or capping layer, is constructed over the slopes and top surface.

Designs for capping layers vary in complexity, just as do those for underliners. The simplest form of capping layer would be a single compacted soil layer of 500 mm or more in thickness which is compacted to a required standard (e.g., 100% of Proctor maximum dry density). A maximum steady-state saturated infiltration rate, established by means of double-ring infiltrometer tests, may be specified to limit the rate of infiltration of rain and the surface should be graded to a minimum slope to shed as much precipitation as possible.

Figure 15 shows an example of a more complex capping layer suitable for a hazardous waste landfill. In this case, because the capping layer incorporates a flexible membrane liner, it should be completely impervious and exclude all infiltrating water from the waste body.

XIV. CONSTRUCTION OF LINERS AND CAPPING LAYERS

The quality and uniformity with which a liner or a capping layer is constructed is most important, and quality must be strictly controlled during construction. The following is a typical specification for the construction of an underliner for a hazardous waste landfill (e.g., see Figure 10):

Prior to the commencement of any construction, the contractor shall submit to the Engineer, for his written approval, the quality control programme for all activities to be carried out on the site. The programme, together with independent checks carried out by the Engineer, shall be sufficient to ensure conformance with the design, specifications and drawings. The correctness of the facility and the quality of the construction must be attested to by the Engineer at the completion of the construction activities.

1. The construction of all leachate containment elements must be supervised on a full-time basis by the Engineer or his delegated representative.
2. The construction of all elements of a Hazardous waste landfill site shall be supervised on a full time basis by the Engineer or his delegated representative.
3. Particular attention must be paid to the quality control of any compacted soil liner.

   The contractor shall carry out a minimum of four density tests per 3000 m² of any compacted 150-mm thick sub-layer. Sufficient Standard Proctor compaction tests must be performed to cover any variability of material that may arise. Density tests using a nuclear device will be considered acceptable provided the results have been proved to be consistent with sand replacement tests. Sand replacement tests will be considered to be the absolute standard for measurements.

   Because the permeability of a soil depends on both the density and the compaction water content, the results of all density tests must meet the following requirements.
   - Dry density equal to or greater than standard Proctor maximum dry density.
   - Water content during compaction within the range:
     Standard Proctor optimum water content to
     Standard Proctor optimum water content +2%.
4. Other earthworks must comply with the requirements of the local standard for earthworks construction.
5. Flexible membrane liners must comply with the requirements of the local standard for flexible membrane liners.

It will be noted that the range of compaction water content is specified to be optimum water content to optimum water content +2%. This is because it has been found that soil compacted within this moisture content range has a lower permeability than if compacted to the same density at a lower moisture content.

Even though a liner has been compacted to Proctor maximum dry density, it may still shrink and crack if allowed to dry out in the sun. For this reason, as soon as any layer of a clay liner has been completed, it should be covered by a minimum 300-mm thick layer of loose uncompacted soil or gravel to prevent it from drying out.

XV. PROOF TESTING OF LINERS AND CAPPING LAYERS

Quite apart from meeting the construction quality control requirements, most regulations for landfills require the liner or capping layer to be proof tested for permeability by means of a field ponding test or double-ring infiltrometer tests.

Field ponding tests are (theoretically) the most reliable method of proof testing a liner because they test a large area of the prototype. However, as described earlier, the tests are difficult to perform accurately because of the problems of accounting for evaporation losses, the effects of variable temperature, etc.

Double-ring infiltrometers, although they test a smaller area of liner or capping layer have a number of advantages:

- A number of tests can be performed simultaneously, thus providing an indication of the variability in quality over a larger area of liner or capping layer.
- The rings can be covered and sealed with plastic sheets, thus eliminating evaporation losses.
- As measurements of flow are made on the inside of the inner ring which is in contact with water on both sides, errors caused by temperature variations are reduced.
- In order to test an adequate area of liner, the inner ring of the double ring infiltrometer should be at least 1 m in diameter, and the outer ring 1.5 m.

Figure 16 shows the results of double-ring infiltrometer tests on a prototype liner. The apparent stepwise decrease in seepage rate that occurs as time progresses illustrates the accuracy with which changes in water level can be measured (i.e., to 1 mm). It will be noted from the time scale (as with Figure 14) that these tests are very time consuming to perform.

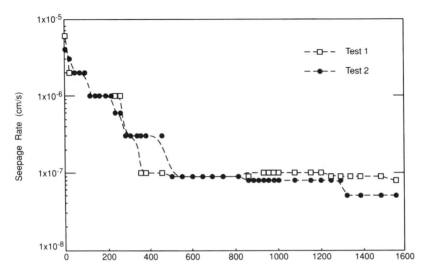

Figure 16 Record of two double-ring infiltrometer proof tests carried out on the lower liner of a hazardous waste landfill.

REFERENCES

1. Blight, G. E. and Blight, J. J., Runoff from landfill surface isolation layers under simulated rainfall, in *Geology and Confinement of Toxic Wastes,* Vol. 1, Arnould, M., Barres, M., and Come, B., Eds., Balkema, Rotterdam, The Netherlands, 1993, 313.
2. Agassi, M., Shainbert, I., and Morin, J., Effect of electrolyte concentration and soil sodicity on infiltration rate and crust formation, *Soil Sci. Am. J.,* 45, 848, 1981.
3. Adamson, P. T., Southern Africa Storm Rainfall. Technical Report No TR102, Department of Water Affairs, Pretoria, South Africa, 1981.
4. Schmidt, E. J. and Schulze, R. E., SCS-Based Design Runoff: Flood Volume and Peak Discharge from Small Catchments in Southern Africa, Water Research Commission Report No. TT 31/87, Pretoria, South Africa, 1987.
5. Ball, J. M. and Blight, G. E., Groundwater pollution downstream of a long established sanitary landfill, in *Proc. Int. Symp. Environ. Geotechnol.,* Fang, H. V., Ed., Envo, Bethlehem, PA, 1986, 149.
6. Blight, G. E., Vorster, K., and Ball, J. M., The design of sanitary landfills to reduce groundwater pollution, in *Proc. Int. Conf. Mining and Industrial Waste Management,* South African Institution of Civil Engineers, Johannesburg, South Africa, 1987, 297.
7. Keenan, J. D., Landfill leachate management, *J. Res. Manage. Technol.,* 14(3), 177, 1986.
8. Saxton, K. E., Soil water hydrology: simulation for water balance computations, in *Proc. Workshop on New Approaches in Water Balance Computations,* Hamburg, Germany, 1983.

9. Fenn, D. G., Hanley, K. J., and De Geare, T. V., Use of the Water Balance Method for Predicting Leachate Generation for Solid Waste Disposal Sites, U.S. Environmental Protection Agency, Report No. EPA/530/SW168, 1975.

10. Burns, J. and Karpinski, G., Water balance method estimates how much leachate a site will produce, *Solid Waste Manage.,* August, 54, 1980.

11. Holmes, R., The water balance method for estimating leachate production from landfill sites, *Solid Wastes,* LXX(1), 20, 1980.

12. Lundgren, T. and Elander, P., Environmental Control in Disposal and Utilization of Combustion Residues, Swedish Geotechnical Institute Report. No. 28E, 1987.

13. Naylor, J. A., Rowland, C. D., Young, C. P., and Barber, C., The Investigation of Landfill Sites, U.K. Water Research Centre, Technical Report TR9, 1978.

14. Campbell, D. J. V., Understanding water balance in landfill sites, *Waste Manage.,* 594, 1983.

15. Blight, G. E., Hojem, D., and Ball, J. M., Production of landfill leachate in water-deficient areas, in *Landfilling of Waste: Leachate,* Christensen, T. H., Cossu, R., and Stegmann, R., Eds., Elsevier, London, 1992, 35.

16. Roper, K. and Fogoqa, F., Determination of the Field Capacity of Domestic Refuse, Final Year Project Report, Department of Civil Engineering, Witwatersrand University, Johannesburg, South Africa, 1988.

17. Blight, G. E., Ball, J. M., and Blight, J. J., Moisture and suction in sanitary landfills in semi-arid areas, *ASCE J. Environ. Eng.,* 118, 865, 1991.

18. Hojem, D., The Water Balance for Landfills, M.Sc. dissertation, Witwatersrand University, Johannesburg, South Africa, 1988.

19. Dullmann, H. and Eisele, B., The analysis of various landfill liners after 10 years exposure to leachate, in *Geology and Confinement of Toxic Wastes,* Vol. 4, Arnould, M., Barres, M., and Come, B., Eds., Balkema, Rotterdam, The Netherlands, 1993, 177.

20. Day, P. R. and Luthin, J. N., Pressure distribution in layered soils during continuous water flow, *Proc. Soil Sci. Soc. Am.,* 17, 87, 1953.

21. Krapac, I. G., Cartwright, K., Panno, S. V., Hensle, B. R., Rehfeldt, K. R., and Herzog, B. L., Water movement through an experimental soil cover, *Waste Manage. Res.,* 9, 196, 1991.

22. Day, S. R. and Daniel, D. E., Hydraulic conductivity of two prototype clay liners, *ASCE J. Geotech. Eng.,* 111, 957, 1985.

23. Chen, H. W. and Yamamoto, L. D., *Permeability Tests for Hazardous Waste Management with Clay Liners. Geotechnical and Geohydrological Aspects of Waste Management,* Lewis Publishers, Chelsea, MI, 1987, 229.

24. Elsbury, B. R., Daniel, D. E., Sraders, G. A., and Anderson, D. C., Lessons learned from compacted earth liners, *ASEC J. Geotech. Eng.,* 116(11), 1641, 1990.

25. Pregl, O., Natural lining materials, in *Proc. Int. Symp. Process, Technology and Environmental Impact of Sanitary Landfill,* Vol. 11, CISA, Cagliari, Sardinia, Italy, 1987, 1.

26. Daniel, D. E., Earthen liners for land disposal facilities, in *Proc. ASCE Speciality Conference on Geotechnical Practice for Waste Disposal,* ASCE, New York, 1987, 21.

CHAPTER 2

Selected Approaches for the Investigation of Microbial Interactions in Landfill Sites

Irene A. Watson-Craik and L. Robin Jones

CONTENTS

0-87371-968-9/95/$0.00+$.50
© 1995 by CRC Press, Inc.

I. INTRODUCTION

Parkes[1] suggested that the main problems in any ecological investigations *in situ* were caused by that very feature that separated the laboratory system from the natural environment, that is, the complex and dynamic nature of the environment. On a gross scale, landfill is an extremely variable and heterogeneous environment, evident in the diversity of refuse composition with both geographical location and time. For example, due to difficulties encountered in measuring the water content of representative samples of refuse used for packing large-scale lysimeters, Campbell[2] had to assign an average value of 25% (w/w) H_2O to samples used. Terashima et al.[3] analyzed refuse composition from four geographical locations in Japan and considered that, for representative analyses, a primary sample of 250 kg was necessary.

Each sample of refuse itself, moreover, constitutes an exceptionally heterogeneous environment which comprises a wide range of organic molecules of both natural and xenobiotic origin, some or all of which may serve as substrates for microbial growth, which are irregularly distributed in a medium composed of surfaces of varying nature and sporadically bathed in a fluid of uncertain composition. Physicochemical parameters such as pH, E_h, a_w, and temperature vary dramatically both spatially and temporally.

Laboratory model systems have the advantage of standardization, reproducibility, ease of control, and manipulation, and Bull,[4] for example, considered that, providing they were made functionally similar to the ecosystem being modeled (e.g., providing appropriate electron acceptors in anaerobic systems), then laboratory systems were sound investigative tools for biodegradation studies. This note of caution was echoed by Slater and Hardman,[5] who considered many laboratory-based ecological studies worthless because insufficient attention had been paid to the properties of the natural habitats, and by Wimpenny et al.,[6] who nevertheless held that, with judicious use, *in situ* and *in vivo* approaches should be complementary, leading at some point in the future to a complete understanding of all the interactions taking place in particular microbial habitats. Slater and Hardman[5] noted that, while the laboratory approach was necessarily reductionist, and significant factors relevant to the system under study were likely to be eliminated by the experimental techniques, it was important that the properties of the original habitat were thoroughly understood and, hence, which parameters were likely to be significant. The complex nature of the landfill ecosystem and its vertical, horizontal, and temporal gradients[7] militates against an easy and complete understanding of its properties. This same heterogeneity is, however, a vitally important property of landfill, for the following reasons:

1. Environmental heterogeneity allows interactions between microorganisms which may not be able to grow together in the same homogeneous habitat.
2. Heterogeneity encourages diversity of microbial types since it provides a wide range of niches for the development of different species.
3. The diverse population is more resilient to environmental stress.
4. The frequency and extent of environmental perturbation and the degree of heterogeneity both determine the rate and range of evolutionary flexibility for particular ecosystems.[6]

As heterogeneity is such an integral property of landfill, heterogeneous models are undoubtedly required to study aspects of landfill biotechnology such as the co-disposal of industrial wastewaters.

II. EXPERIMENTAL MODEL SYSTEMS

Landfill is an example of a habitat where the microflora are located in a fixed position and are subject to the flow of solutions. Investigations on similar habitats such as soils, tidal rocks or surfaces, and rocks on stream and river beds have concentrated on the use of percolating columns,[6] where downward nutrient flow leads to vertical stratification of microorganisms and their metabolites. These columns may employ natural packings, such as soil and sediments, or synthetic materials such as glass beads.[8] While the use of inert packings such as glass beads has significant advantages in simplification of microbial behavior, the range of surfaces in landfill provides not only a matrix for microbial attachment but also the substrate itself. Similarly, relevant studies of landfill model systems require the enrichment of an indigenous mixed population of interacting bacterial species.

The designs of landfill model systems reported have ranged from simple batch cultures[9] through test columns (0.75 m³) with leachate recycle[10] to pilot-scale lysimeters (8 m³) with leachate recycle.[11]

A. CLOSED CULTURE MODELS

The biochemical methane potential (BMP) assay, a batch method typically used to assess anaerobic biodegradability of liquid wastes with added nutrients and bacteria,[12] was adapted[13] to study biodegradation rates of landfill samples. Ground (2 mm) and homogenized refuse samples were dispensed (25 g) into serum bottles, which were overgassed with a nitrogen:carbon dioxide (70:30) mix, sealed, and incubated at 35°C. Biodegradability was assessed in terms of methane production. It was suggested that this method could be adapted for biodegradability and anaerobic toxicity studies for potential landfill co-disposal of hazardous wastes, decomposition of selected refuse components such as "biodegradable" plastics, or to study the effects of environmental parameters such as moisture content. Bogner[13] demonstrated, by use of this method, that anaerobic sampling methods had no stimulatory effect on the onset of methanogenesis from refuse samples (Figure 1). However, this approach provides little insight into the nature of microbial interspecies interactions, and can only suggest, by inference, the specific targets of, for example, inhibitory or toxic molecules.

This was demonstrated by Watson-Craik et al.,[14] who reported the inhibitory effects of o-cresol on the development of methanogenesis in batch refuse cultures. When a range of o-cresol concentrations (0 to 7 mM) was added to refuse samples on day 1, significant inhibition of methanogenesis was apparent at o-cresol concentrations $\geq$4 mM (Figure 2a). However, when the cultures were supplemented with the xenobiotic on day 27, no significant inhibition was recorded at concentrations $\leq$7 mM (Figure 2b). It was proposed that the methanogens were not directly inhibited by these concentrations of o-cresol but that other populations, possibly sulfate-reducing bacteria (SRB), on whose activities the development of an active methanogenic populations depended, were inhibited by cresol concentrations $\geq$4 mM. However, this could only be inferred, not directly demonstrated, by use of these simple laboratory model systems.

Methods for the evaluation of leachability of a test chemical often employ a batch method. These usually include a synthetic leachate such as 0.05 M citrate buffer[15] or acetic acid (5 g/l) buffered at pH 5.84. Such methods ignore both the contribution of biological activity to leaching of chemicals and the influence of flow. Anaerobiosis has, for example, been found to promote leaching of metals such as nickel, chromium, zinc, lead, and cobalt,[16] while solute flow may transport, for example, acids that increase the leaching of lead.[17] Jackson et al.[18] compared column and batch methods for assessing leachability and concluded that, while the latter were simpler and showed greater reproducibility, column methods were more realistic.

B. REFUSE COLUMNS

Refuse columns (2-l capacity) were used by Barlaz et al.[19] to study the effects of leachate recycling and seeding on the microbial, chemical, and methane production characteristics of pulverized refuse. It was concluded that the total anaerobic population and the individual cellulolytic, acetogenic, and acetoclastic and H_2/CO_2-utilizing methanogenic populations were the same whether the refuse was incubated with leachate recycle at 72% moisture or without leachate recycle at 48% moisture. In those columns seeded with anaerobically decomposed refuse, sulfate concentrations were higher (72.5 to 258 mg/l) than those recorded in columns packed with

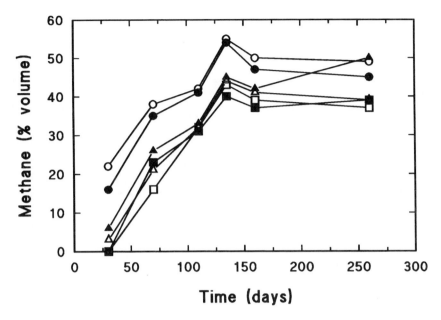

Figure 1   Cumulative headspace methane concentrations in sealed batch cultures inoculated with refuse sampled aerobically (○, □, △) or anaerobically (●, ■, ▲). Refuse samples were incubated with no moisture addition (□ and ■), with the addition of distilled water to equal gravimetric moisture content of 200% (△ and ▲), or with the addition of nutrient medium to equal gravimetric moisture content of 200% (○ and ●). (Adapted from Bogner, J. E., *Waste Manage. Res.*, 8, 329, 1990).

fresh refuse only and operated with leachate recycle (0.5 to 6.9 mg/l). However, it was not clear whether this could be directly attributed to differences in the inocula or to interactions between the populations.

Small-scale refuse columns (1-l working volume) were also used in a series of studies[20–22] to study the treatment of phenolic wastewaters by co-disposal with domestic refuse and the effects of the co-disposal of phenolic wastewaters both on the development of the refuse methanogenic fermentation and the microbial interspecies interactions. Interest in this was stimulated by the persistence of high concentrations (8 to 375 mg/l) in the leachates and boreholes of several U.K. landfill sites,[23] which suggested that phenol degradation was slow.

Methanogenic fermentation of phenol has, however, been reported with inocula from anoxic aquifers[24,25] and anaerobic digesters.[26–28] It may be speculated that phenol plays a pivotal role in the methanogenic fermentation of a wide range of naturally occurring aromatic substrates since lignin monomers such as vanillate, ferulate, protocatechuate, and syringate have been shown to be dissimilated to catechol as an intermediate[29,30] before oxidation to phenol, through *cis*-benzenediol.[29] *p*-Cresol and phenol have

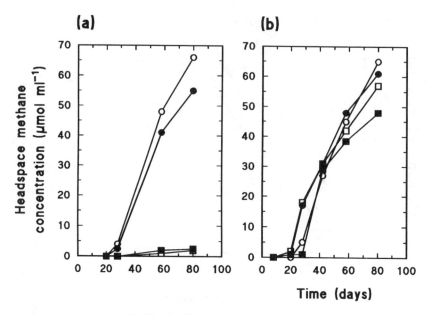

Figure 2   Cumulative headspace methane concentrations in batch cultures supplemented with 0 (o), 3.5 (●), 4 (□), or 7 (■) mM o-cresol on (a) day 1 or (b) day 27. (Adapted from Watson-Craik, I. A., Sinclair, K. J., James, A. G., Sulisti, and Senior, E., *Water Sci. Technol.*, 27, 15, 1993.)

also been detected as intermediates in the methanogenic fermentation of the naturally occurring amino acid tyrosine.[31] Since aromatic acids, such as vanillic, ferulic, and syringic, and amino acids, such as tyrosine, are likely to be widespread in waterlogged soils, sediments, and refuse, even if localized and at very low concentrations, it seems likely that the capability exists in the genetic pool for the degradation of phenol to CO_2 and CH_4. Suboptimal degradation in landfills may be due, therefore, to excessive loading rates or to limiting environmental conditions.

Single-stage refuse columns, operated with either leachate recycle or leachate discard, were perfused with a range of phenol concentrations (0 to 12 mM), and the development of a methanogenic fermentation was monitored.[21] In columns operated with leachate recycle, decreased leachate acetate concentrations were only recorded, on day 23, with influent phenol concentrations ≥8 mM (Table 1). With leachate discard, inhibition of volatile fatty acid (VFA) production/release was apparent at phenol concentrations ≥1 mM. Methane assays confirmed that the low VFA concentrations could not be attributed to stimulation of methanogenesis. Indeed, the low methane (0.17%) and increased acetate (14.5 mM) concentrations recorded in the leachate recycle column perfused with 4 mM phenol suggested that methanogens may be more susceptible than acetogens to the inhibitory effects of phenol.

Table 1   Changes in leachate acetate and headspace methane concentrations in refuse columns operated with either leachate discard or recycle and perfused with a range of phenol concentrations

Perfusion strategy	Influent phenol concentration (mM)	Leachate acetate concentration (mM)	Headspace methane concentration (%)
Leachate discard	0 (control)	13.4	0.38
	1	2.1	0.06
	2	6.3	0.10
	4	0.9	0.01
	8	0.8	0.02
	12	0.5	0.01
Leachate recycle	0 (control)	12.9	0.49
	1	10.2	0.59
	2	9.9	0.85
	4	14.5	0.17
	8	6.5	0.02
	12	1.7	0.01

Adapted from Watson-Craik, I. A. and Senior, E., *Water Res.*, 23, 1293, 1989.

The effects of elevated phenol concentrations on the different physiological groups involved in the methanogenic fermentation of refuse organics and intermediate products were also observed in a multistage refuse column array.[22] In this study, eight glass columns each packed with 1 kg uniformly compacted refuse were serially linked (Figure 3) and a model phenolic wastewater was serially perfused through the array at an empty bed dilution rate (D) of 0.026, 0.013, 0.007, 0.004, and 0.003/h at sampling ports 1 to 5, respectively. After 45 days perfusion with 2 mM phenol, no phenol was detected at sampling ports 3, 4, or 5, while residual concentrations of leachate samples 2 and 1 approximated to 14 and 32%, respectively, of the influent. Increases in concentration, to 4 and 5 mM phenol, had little effect, with no residual phenol recorded at sampling ports 4 or 5. However, when the concentration was increased to 8 mM, after 45 days perfusion, residual phenol was recorded at ports 4 (0.6 mM) and 5 (0.1 mM). These residual concentrations then progressively increased over 49 days to 6.5, 5, and 3.3 mM at ports 3, 4, and 5, respectively. Reduced phenol catabolism was not reflected by low leachate pH values or high VFA concentrations. This suggested that either the phenol concentration applied inhibited the acidogenic species also or the acidogenic substrate pool was depleted. The rates of methane release were not unduly affected. Since hexanoic acid has been suggested as an intermediate in the anaerobic degradation of phenol in the presence of nitrate reduction[32] and methanogenesis,[33] it was questioned whether 8 mM phenol would inhibit its dissimilation. The influent was, therefore, supplemented with 2 mM hexanoate in PO_4 buffer (4 mM). Within 8 days, the hexanoate was completely dissimilated (D = 0.007/h), with simultaneously increased methane release, although residual phenol concentrations increased over the same period. Since anaerobic hexanoate degradation was not constrained either by nutrient limitation or the

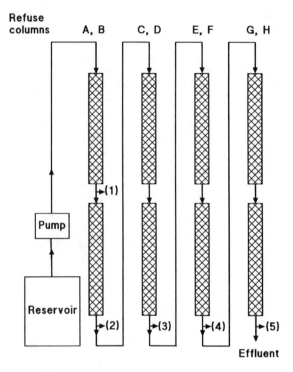

Figure 3   Multi-stage refuse column array (columns A to H), with sampling ports (1) to (5). (Adapted from Watson-Craik, I. A. and Senior, E., *J. Chem. Technol. Biotechnol.*, 47, 219, 1990.)

potentially toxic/inhibitory effects of the perfused phenol, it appeared that the limited attenuation of phenol was due to inhibition of phenol transformation to hexanoate. A similar differential inhibitory effect was reported by Fedorak and Hrudey,[34] who showed that in batch culture acetogenesis and methanogenesis were inhibited by phenol concentrations >21.5 mM in contrast with the phenol-degrading acid formers that were inhibited by concentrations ≥8.5 mM.

Both single- and multi-stage models were, however, limited in their applicability to the elucidation of microbial interspecies interactions since the nature of the interactions could only be inferred rather than directly demonstrated.

The environmental relevance of some column studies can also be questioned. Rees and King,[35] for example, modeled anaerobic phenol degradation in the unsaturated zone below landfill sites but employed a synthetic leachate based on acetate (8400 mg/l), propionate (3000 mg/l), and mineral salts, which included PO_4-P (113 mg/l) and SO_4 (800 mg/l). Leachate, however, is particularly variable. Thus, for example, Cheyney[36] recorded 242 different compounds in the leachate from one site and Robinson and

Maris,[37] who analyzed leachate from 16 sites, reported concentrations of acetate, propionate, ortho-P, and SO_4 which ranged from below detection limits to 2805, 1106, 4.4, and 456 mg/l, respectively. The elevated concentrations used by Rees and King[35] may thus have induced effects that were not relevant to *in situ* activity. For example, increased inhibition of activity by phenol has been reported[38] at volatile fatty concentrations > 500 mg/l.

The importance of relevant feeding regimes and conditions was emphasized in the studies of Gourdon et al.,[39] who examined n-butyrate catabolism in the treatment of a semi-synthetic landfill leachate, using an anaerobic trickling filter packed with vermiculite. Under conditions of continuous feeding, butyrate was β-oxidized to acetate. However, under sequential feeding conditions, *iso*-butyrate was formed at about 25% of the total amount of n-butyrate removed and, subsequently, decarboxylated to propionate. Catabolism of propionate to acetate and CO_2, and subsequent methanogenesis of acetate to CO_2 and CH_4, resulted in the production of 1 mol of methane per mole butyrate, compared to the 2 mol methane which resulted from β-oxidation. It is clear, therefore, that feeding regimes may have significant effects on observed methane release rates.

The composition of other influent components is also important. Stanforth et al.[40] described the development of a synthetic leachate, modeled after data available from on-site investigations, but admitted that paucity of information limited the development of a representative synthetic leachate or indeed a range of leachates to model a range of site conditions.

Fungaroli and Steiner[41] proposed a list of specifications that should be satisfied for the simulation of landfill behavior. They considered, for example, that

1. The heat exchange across the vertical walls of the bioreactor must be at a minimum compared to heat exchange at the boundary equivalent to the atmosphere-soil cover interface.
2. The refuse temperature must be equivalent to the average annual *in situ* soil temperature at the same depth.
3. The tank cross-sectional area must be large enough to ensure the generation of reliable experimental data.
4. The refuse composition must be representative.
5. Individual refuse component sizes must be selected to produce a lysimeter behavior equivalent to a full-size landfill.
6. The percent H_2O content at placement must be the natural value.
7. Refuse densities at placement must be in the range usually found in landfills.

Although large-scale lysimeters satisfy the first two criteria better than small-scale columns and, although their ability to predict leachate reactions is good,[42] their capital cost is high, considerable operator input is required, and the speed of obtaining results is slow. Moreover, there may be persistent difficulties in maintaining anaerobiosis[41] due to the methods required for lysimeter construction.

Small-scale columns offer the advantages of low capital cost, rapid results, and lower operator input,[42] although the design of environmentally relevant model column systems should consider both the criteria proposed by Fungaroli and Steiner[41] and present understanding of landfill behavior. In addition, by the judicious use of control columns, the behavior of the system should be continually assessed and compared with published evidence of *in situ* landfill behavior such as, for example, temporal leachate compositional changes. Nonetheless, the validity of extrapolation of laboratory-based results back to the natural habitat remains a central problem.

Few landfill studies have addressed the problem although Newton,[43] for example, compared leaching from both concrete tanks (cross-sectional area 5 m^2) and PVC pipes (cross sectional area 0.071 m^2) when cyanide, metal hydroxide sludges, or oil emulsion were added to domestic refuse, and reported that the pattern of leaching of degradation products did not differ significantly. The efficacy of leachate recycle as a treatment option for leachate was investigated both in pilot-[11] and field-scale studies.[44] From the results it was concluded that, while leachate strength was significantly reduced in both treatments, it was more difficult to reduce the field-scale leachate to the lower concentrations obtained in the pilot-scale studies.

The limitations of small-scale model column systems for mathematical and kinetic analyses have, however, been demonstrated.[45] The assumption, for example, that a column is at a steady state can lead to errors. If it is not, then parameters of time and distance down a column must be taken into account. From Figure 4 it can be seen that the specific reaction rates of substrates and microbial biomass are dependent on time and on spatial position within the column. However, quasi-steady states have been reported, with low input concentrations.[46]

C. CONTINUOUS CULTURE MODELS

The homogeneity of chemostat operation, where the microorganisms are in a perfectly mixed suspension into which the sterile medium is fed at a constant rate and the culture is removed at the same rate, leads to the development of a steady state, or to the approximation of one ("stable state") for mixed cultures. Such a state can be considered to be time independent since parameters such as culture conditions and biomass concentrations are constant. With time independence a system lends itself to mathematical analysis and kinetic parameters can be determined by measuring steady-state kinetics at a variety of dilution rates and substrate concentrations.

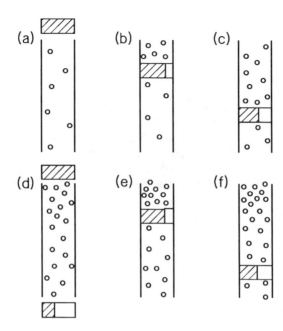

Figure 4   Schematic representation of microbial degradation of a single substrate in a column packed with an inert support and perfused dropwise. Temporal changes (a to f) in the density of the biomass are represented by the density of the dots, and the concentration of substrate by the length of the shaded portion. (Adapted from Bazin, M. J., Saunders, P. T., and Prosser, J. I., *Crit. Rev. Microbiol.*, 4, 463, 1976.)

The continuous culture study approach may, therefore, be selected to investigate some of the parameters which affect the degradation of constitutive and xenobiotic substrates by microbial associations enriched and isolated from landfill. Two-stage systems have been used extensively[47–49] to investigate the anaerobic digestion of carbohydrate-containing wastewaters or to model the anaerobic degradation of cellulosic materials. Girard et al.,[47] for example, localized, in the first stage, organic acid production from cellulose [0.5 to 2.3% (w/v)] at a C:N ratio of 4:1. High loading rates favored butyric acid production while at low loading rates (0.5g/l/d) acetic and propionic acid were predominant. Methanogenesis from the organic acid mixture was localized in the second reactor. However, these systems do not adequately model the more complex and heterogeneous landfill ecosystem, with temporal fluxes in a range of electron acceptors such as nitrate, sulfate, and carbon dioxide.

Multi-stage systems have, therefore, been developed, in which the effects of a range of electron acceptors at environmentally relevant concentrations can be studied. Parkes and Senior[7] considered multistage chemostats ideal models for studying anoxic ecosystems since anaerobic decomposition proceeds via the sequential intervention of different physiological types and there is, thus, the potential for isolating each sequential event within individual vessels of the array. As these are open systems, distribution of the various physiological types within them is time independent and, therefore, mathematical and kinetic analyses may be applied. Both five- and three-stage chemostats were employed[50] for the elucidation of the interspecies interactions within a microbial association, enriched and isolated from domestic refuse, effecting terminal anaerobic catabolism of hexanoic acid, a representative low molecular weight molecule and key intermediate found in leachate of recently emplaced refuse.[51] Partition of physiological groups reducing the electron acceptors nitrate, sulfate, and carbon dioxide was demonstrated by use of a four-stage model.[14] At low dilution rates (0.032/h in the first vessel), all nitrate-reducing, sulfate-reducing, and methanogenic activity was localized in the first vessel (Figure 5). However, at imposed dilution rates which ranged from 0.247 to 0.050/h in the first to the fourth vessel, nitrate reduction was retained in the first vessel, while sulfate reduction was partially displaced, with 67 and 90% cumulative removal in the first and second vessels, respectively. The slower-growing methanogens were simultaneously displaced into the third and fourth vessels (Figure 5).

Once such separation has been achieved, species response to shock loads and overloads, inhibitory molecules, or changes in concentrations of electron acceptors such as nitrate may be examined.

For example, the effects of temperature on the methanogenic fermentation of the cellulose-degradation intermediate cellobiose, by an association enriched and isolated from anoxic refuse, was studied by use of a three-stage system.[52] The association stabilized at dilution rates of 0.076, 0.030, and 0.020/h in vessels A, B, and C, respectively, at which substrate attenuation and sulfate reduction were maintained predominantly in the top vessels. However, more than 90% of methanogenic activity was displaced from vessel A to vessels B and C, presumably reflecting the low μ_{max} values of methanogens compared to cellobiose-degraders and SRB. The main observed intermediates of cellobiose catabolism were acetic, propionic, and butyric acids. Vessels A and B were characterized by extensive acidogenic and acetogenic populations while, in vessels C, some acetogenic (propionate-degrading) and approximately 10% of the overall methanogenic activity were located.

In all vessels, the optimum temperature range for methanogenesis was 30 to 35°C (Figure 6a and b). Methane release rates were inhibited at ≥40°C, and no thermophilic activity was evident. However, in vessels B and C, methanogenic activity was not inhibited at temperatures <30°C, suggesting that vessels B and C may be characterized by different types of methanogens from vessel A. Net acetate concentrations varied little (5 to 7 mM) over the temperature range 20 to 55°C, reflecting a fairly

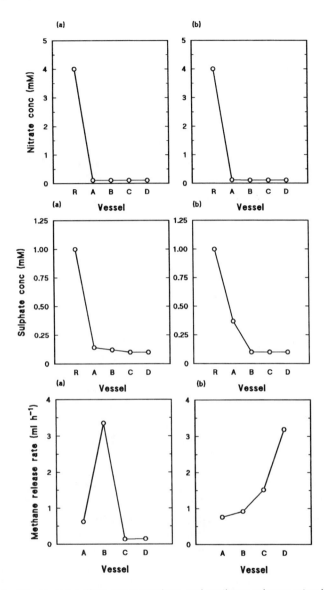

Figure 5   Nitrate and sulfate concentrations and methane release rates in a four-
stage continuous culture model system, in which vessels A, B, C, and D
were operated at dilution rates (a) 0.032, 0.020, 0.012, and 0.006/h and
(b) 0.247, 0.157, 0.100, and 0.050/h. R denotes the concentration of nitrate
or sulfate in the reservoir. (Adapted from Watson-Craik, I. A., Sinclair,
K. J., James, A. G., Sulisti, and Senior, E., *Water Sci. Technol.*, 27, 15,
1993.)

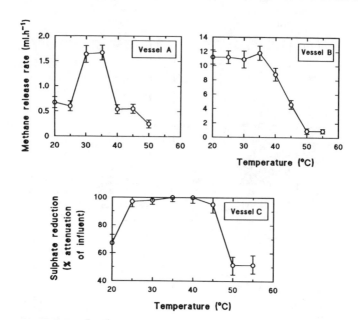

Figure 6   Effects of temperature on methane release rates in vessels A and B, and
sulfate reduction in vessel A, of a three-stage system, inoculated with an
anaerobic cellobiose-degrading association. Methane release rates in
vessel C displayed a similar response to those in vessel B. (Adapted from
Watson-Craik, I. A., James, A. G., and Senior, E., Proceedings of the 7th
International Symposium on Anaerobic Digestion, Cape Town, 1994, 2.)

constant balance between genesis and trophy. Inhibition of butyrate and
propionate degradation above, respectively, 45 and 35°C, was partially attributed to inhibition of the hydrogen-sink bacteria.

In vessel A, SRB activity varied little over the temperature range 25 to
45°C (Figure 6c). However, although inhibited at 50°C, an increase to 55°C
had no further effect. These data suggested that the cellobiose-degrading
association was characterized by thermophilic, possibly propionate-
utilizing, sulfate reducers. Both net butyrate and, as in vessel B, net acetate
concentrations varied little (0.8 to 1.5 and 8 to 10 mM, respectively) over
the temperature range 20 to 55°C, reflecting a fairly constant balance between genesis and trophy. Reduced propionate concentrations were, however, evident at 20 and above 35°C, probably due to a combination of
inhibited propionogenesis and, at the higher temperatures, the action of
thermophilic SRB. No inhibition of the cellobiose-degrading bacteria was
evident over the range 20 to 55°C. These responses to environmental perturbation indicated that associations enriched from refuse may comprise
several types of methanogens and SRB. Methanogenic populations were
inhibited at higher temperatures (45 to 55°C), and prolonged exposure to

these temperatures *in situ* may result in an unbalanced fermentation or the redirection of electron flow to thermophilic, propionate-utilizing SRB.

Although the temporal and spatial heterogeneity of landfill temperature is well known,[53] refuse emplacement strategies have been described as a means of controlling the temperature within a landfill site. Rees,[54] for example, suggested that low-density refuse, diluted with wood, paper, pulverized, or aerobically stabilized refuse, should be placed at the landfill base and allowed to decompose. Exothermic aerobic catabolism would then effect a temperature increase and heat subsequent layers of refuse, with heat loss from the site minimized by an insulating layer of refuse above the reactive zone. There has, however, been little research on the viability of operating landfills within the narrow temperature range (30 to 35°C) shown in this study to be optimal.

The use of continuous culture studies for the acquisition of kinetic data and the derivation of realistic loading rates for the co-disposal of specific organic wastewaters has not been extensively reported. One series of studies[14,55,56] examined the effects of the co-disposal of phenolic wastewaters on the fermentation of hexanoate, an organic acid characteristic of leachate,[51] by a methanogenic association isolated from anoxic refuse. A three-stage continuous culture model was used to segregate the hydrogen-sink bacteria, such that the methanogenic population was contained in vessels B and C (D = 0.04 and 0.02/h, respectively) and the SRB in vessel A (D = 0.10/h). Supplementation of vessel C with ≥ 8 mM phenol resulted in significantly decreased methane release rates, from 1.54 ml/h (4 mM phenol) to 0.15 ml/h (8 mM phenol). Hexanoate dissimilation, located in vessels A and B, was not affected when challenged with ≤ 8 mM phenol. With influent 8 mM phenol, residual acetate concentrations in vessel B did not decrease which, coincident with a methane release rate decrease from 3.58 (4 mM phenol) to 2.26 ml/h (8 mM phenol), suggested that acetotrophic methanogenesis was at least partially inhibited by 8 mM phenol.

When vessel A was supplemented with 2 mM phenol, the concentrations of both residual hexanoate and sulfate significantly (99% confidence limits) increased. Supplementation of vessel B with the same phenol concentration had demonstrated no inhibition of hexanoate dissimilation, with the methanogenic population presumably providing the electron sink for the generated hydrogen. In vessel A, however, it appeared that the major sink for hydrogen was provided by the SRB and that inhibition of their activity resulted in partial inhibition of hexanoate degradation.

It is clear that continuous culture multistage systems can yield useful information on anaerobic microbial interspecies interactions, including the effects of environmental parameters and added xenobiotics. However, in these models the scope for kinetic analyses proved limited; this was due partly to the design of the perfusion system and the resultant operation on at least a partial plug flow basis, and to the presence of wall growth in all vessels despite silanization of the culture vessels prior to operation.

Surface attachment and growth is likely to be a significant factor in microbial activity and interactions in landfill sites: refuse is characterized by materials that may serve not only as organic substrates for growth, such as paper and plant cuttings, but also as supports for microbial biofilms. Other materials, such as glass, plastics, and metals, usually thought of as microbially inert, may play an important role in biofilm maintenance and, hence, microbial activity in landfill sites.

A method was, therefore, required which could be used to investigate the real time processes of attachment, establishment, division, and growth of anaerobic microorganisms *in situ,* on surfaces characteristic of domestic refuse.

D. THE ANAEROBIC CONTINUOUS CULTURE MICROSCOPY UNIT (ACCMU)

The ACCMU was developed to model and observe the development of anaerobic microbial biofilms on materials characteristic of refuse emplaced in landfill sites.

The device is a miniature continuous culture unit that fits onto a standard microscope stage. It is incubated inside an anaerobic cabinet and is self-sealing when detached and removed for microscopic observation. Though the application of such an approach to the study of biofilms in landfill sites is new, the principle can be traced as far back as Pasteur.[57] "Modern" units for the observation of living microorganisms, known as perfusion chambers, date from over 40 years ago and many have been described by Quesnel.[58]

They include the "Pomerat Chamber", made from two coverslips held together by wax, in a metal block; the "Rose Chamber", made of two steel plates with a central hole, separated by latex, with coverslips between the plates and latex; and the "Toy and Bardawil Chamber", which consisted of three plexiglass plates which held two coverslips apart. Medium entered and left through two ducts in the top plate. The "Ware and Loveless Chamber" was formed from a coverslip separated from a microscope slide by glass strips, with glass cylinders at each end. Time-lapse photography was used to keep a record of the development of sewage biofilms. Quesnel[58] described the "Carter Chamber" as "beautifully simple and efficient"; two concentric stainless steel rings, bonded to glass and separated by an annular channel were sealed by filling the outer ring with heavy silicon oil and fitting a coverslip above with spring clips. Access was by a steel tube and the contents could be agitated electromagnetically.

A feature of the "Rose Chamber", and some later ones, was that influent medium was fed in by hypodermic needles, which could be inserted or removed easily, a principle that was applied in the development of the ACCMU.

Harris and Powell[59] described a stainless steel culture chamber, fitted into an aluminum carrier, that attached to a microscope stage, and which was used to study colonization of cellophane. An improved "Powell Chamber" was later built. The new "Powell Chamber" was described by Quesnel,[58] who stated that it "opened up a whole new field of microscopic observation." The "Powell Chamber" was complex, utilizing interconnecting parts which required precision engineering.

Duxbury[60] devised an improved, simpler, stainless steel "Microperfusion Chamber" for the study of cellophane surfaces. It was fed by hypodermic needles and incorporated temperature control. Its three-piece, circular construction was designed to fit different microscope stages, depending on the mode of assembly. Berg and Block[61] described a similar "miniature flow cell", also of stainless steel, with integral supply tubing and glass coverslips fitted to the top and base.

Many of these devices, while undoubtedly sophisticated, required accurate engineering and assembly. The investment in time, and the cost, meant that they could not be regarded as disposable items, and cleaning them for reuse was a further investment in time.[61] Sjollema et al.[62] used a parallel plate flow cell, with a removable upper portion for cleaning, but the unit still required accurate machining of parts.

Perfilev and Gabe[57] provided comprehensive descriptions of the various ways in which capillaries could be used to study microorganisms. They adopted a simpler approach with square, glass capillaries through which growth medium was pumped. Biofilm development on the capillary wall was observed microscopically. Their method was used, with the addition of temperature control, by Rutter and Leech[63] to study surface accumulation of *Streptococcus sanguis*.

Glass capillaries often contain optical imperfections due to the small distance between vertical walls and stretching marks resulting from the manufacturing process.[64] An alternative method is to make continuous culture units, or flow cells, from microscope slides and coverslips. The operation and construction of these units was first described by Caldwell and Lawrence.[64,65] The unit was constructed from two microscope slides side by side, 2 mm apart, with a coverslip fixed with silicon rubber adhesive centrally across the slides, above and below, creating an enclosed chamber 2 mm × 1 mm in section and the length of the coverslips. Tubing was fixed to the ends of the chamber for medium supply. The unit was cheap, disposable, and fit onto a standard microscope stage, and is the design on which the ACCMU was based.

In recent years the techniques of continuous observation of development, described above, have been used together with computer-enhanced image analysis of phase contrast and darkfield microscope images, to study the number of cells on surfaces, the initial stages of cell attachment, and the division of single attached cells to form colonies.[62,66,67] The availability of powerful, user-friendly image analysis systems has greatly improved the speed and accuracy of discrimination and quantification of cells growing on surfaces.[68,69] Colonization, growth, and behavior of aerobic bacteria on glass surfaces have been described and quantified.[64–67,70] Such techniques have allowed the rate of aerobic growth of single bacterial strains on surfaces to be quantified in terms of the increase in cell area and cell number with time and interdivision time.[65]

1. ACCMU Construction and Operation

The flow cell device and image analysis, when combined with anaerobic continuous culture, indicated an approach by which biofilm development in landfills might be modeled and studied. Previously, the flow cell had been used for the study of aerobic microorganisms, the rapid growth rate of which allowed short experimental durations, with the flow cell attached to the microscope stage throughout. The slow growth rates of anaerobic microorganisms required extended incubation periods, with a consequent risk of oxygen contamination of bench sited systems. Conducting microscopy inside an anaerobic cabinet was considered impractical.

The adopted approach of a detachable unit, incubated anaerobically and removed for microscopy, held the advantage that, as the unit was not permanently positioned on the microscope stage, the microscope was available for other uses between samples. It also meant that, with the aid of a multichannel peristaltic pump, more than one unit could be incubated and sampled at a time. The ACCMU was built from two microscope slides (2 mm apart), two coverslips (64 mm × 30 mm) and two butyl rubber plugs, all secured with silicon rubber adhesive (Figure 7). The whole system was maintained in an anaerobic atmosphere of 80% N_2, 10% H_2, and 10% CO_2 inside an anaerobic cabinet at 30°C.

a. Inoculation Procedure

Inside the anaerobic cabinet, the ACCMU was fitted with a hypodermic needle at one end and cabinet atmosphere was injected from the other end. This was to remove any residual oxygen from the assembly. Butyrate- and hexanoate-degrading associations, enriched from municipal solid waste (MSW) under batch culture conditions,[71] were used as the inocula. The inoculum was introduced into the outflow end of the ACCMU with a hypodermic needle, to 0.25% of the chamber volume (approximately 0.03ml), and the ACCMU left in the cabinet for at a minimum of 12 h.

(a)

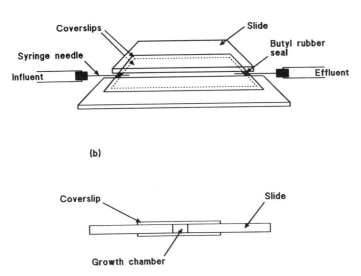

(b)

Figure 7   (a) Plan and (b) end view of the ACCMU, constructed of two microscope slides, two long coverslips, and two butyl rubber plugs.

b. Continuous Culture System

Hypodermic needles (25 G) attached to tubing (Marprene 1.65 mm i.d.) were inserted through the butyl rubber and supplied medium from a reservoir, driven by a peristaltic pump at a rate of 1.5 to 2 ml/h. Thus, calculated bulk flow velocity in the ACCMU was 75 to 100 cm/h. The combination of an eight-roller pumphead and small bore manifold tubing provided a series of small pulses, approximating to a uniform flow. It was important to avoid shear stress effects on the developing biofilm due to variations and reversals in flow often associated with peristaltic pump operation.

Gas bubbles can also cause removal of cells but can be blocked by inserting a bacterial filter before the input.[67] In addition, a pre-filter bubble trap, in the form of an adapted 1-ml syringe barrel, can be fit vertically into the filter inflow (Figure 8).

The culture medium was complete mineral salts, supplemented with butyric or hexanoic acid (5 mM) as carbon source.[72] These VFAs were chosen as carbon sources since they are important intermediates of the anaerobic degradative pathways leading to methanogenesis.[51]

The reservoir and tubing were autoclaved (121°C, 1 bar, 15 min) prior to use, and the ACCMU chamber was sterilized with ethanol (100%) at assembly. The medium was filter sterilized.

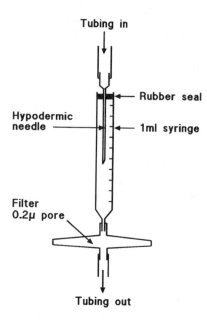

Figure 8 Diagram of a bubble trap made from a 1-ml syringe and a hypodermic needle. Gas emerging from the needle rises to the top of the syringe barrel and so is not passed through the filter to the ACCMU.

c. Sampling

At sampling times, the ACCMU was disconnected from the medium supply by withdrawing the hypodermic needles and removed from the anaerobic cabinet. The ACCMU was placed on the stage of a microscope (Nikon Optiphot II) and colonization of the underside of the top coverslip observed using phase-contrast illumination. In addition, methanogenic bacteria could be located by their characteristic blue-green fluorescence with an excitation wavelength of 420 nm (Nikon BV-2A filter block).[73] Both phase contrast and fluorescence images of the same field could be obtained by switching between the two methods of illumination.

A CCD video camera (Sony, 90C) fed microscope images to an image analysis computer (Seescan Cambridge, Solitaire) and, in this way, microscope fields were captured and stored on hard disk. The X-Y coordinates on the stage vernier scale were recorded for each field. After the capture of fields, the ACCMU was returned to continuous culture mode in the anaerobic cabinet.

At subsequent sampling times, the fields were located using the stage coordinates and accurately aligned with previous images using the Seescan "split" program, which permits the simultaneous display of a previously stored image with the current live image. In this way a time sequence of the microbial colonization was obtained for each field and was stored on disk.

Where photofading of the fluorescence was rapid, less than 1 s, the fluorescence image was best captured first, using the lowest possible level of phase illumination for field alignment only. Fluorescence illumination was then applied only long enough for image capture. This minimized both photofading and exposure of the culture to the potentially harmful short wavelengths used. Further reduction of exposure could be achieved by restricting, where possible, the capture of fluorescence images to the end of the experiment. After photofading, the fluorescence returns if the excitation illumination is removed for a few minutes, and this provides another way to "build up" fluorescence images where photofading is rapid.

d. Image Analysis

The use of image analysis for this work has several important advantages:

1. Since analysis, such as the measurement of cell parameters, e.g., length, breadth, area, perimeter length, is done on the stored images, the time that the ACCMU is exposed to microscope illumination—only required for locating fields and capturing images—is reduced. Thus, possible effects of light on the microorganisms[74] are minimized. Where the fluorescence is too faint to be detected by a standard image capture program, it may be captured using an image analysis program (averaging) which increases sensitivity by summing a rapidly captured sequence of images.
2. The length of time that the ACCMU is exposed to atmospheric oxygen is minimized, reducing the chance of oxygen contamination of the culture.
3. The length of time that the ACCMU is removed from continuous culture is also minimized.
4. The analysis does not have to be done at the time of sampling and so can be done later, at a more convenient time. In this way, image analysis can facilitate efficient use of time, and the need to observe the culture for long periods under the microscope is avoided.
5. Stored images can be returned to for new analyses, after the original experiment has ended. This can be important as new insights may be gained from initial analyses and as new and more sophisticated image analysis programs become available.
6. The image analysis computer greatly speeds the analysis process. For example, up to eight measurements can be made on all the "thresholded" objects on the screen in a single operation. Up to ten calibrations, e.g., for different objective-zoom combinations, can be stored and recalled. The data are stored in a results file and standard programs provide rapid presentation of basic statistics and histogram and scatter diagram plots of chosen parameters.

The speed at which images can be processed, and the accuracy of image analysis, is dependent upon consistent "thresholding", such that measurements are reproducible and comparable. Thresholding is the discrimination of cells from the background and from other objects. It relies upon differences in light intensity and can cause problems where light intensity and contrast is low, e.g., at high magnifications. The accumulation of debris on the surface also makes thresholding more difficult by creating artifacts of similar intensity to the cells and obscuring their outlines. However, even in a worst-case situation, where cells recognized by the operator have to be individually counted or measured, the process is as rapid as alternative methods. Where the images are amenable to thresholding, image analysis is capable of acquiring rapidly data which by other methods requires great effort and time.

Digitized images use a large amount of computer memory. Therefore, in order to gain the potential benefits of long-term image storage, it is essential to have an archiving system such as a WORM (write once read many) drive. Archiving allows the images on the hard disk of the image analysis computer to be off loaded in order to provide space for storing the images from the next experiment. Each WORM cartridge has the capacity to store thousands of images and, as with floppy disks, there is no limit to the number of cartridges that can be obtained. An alternative is laser disk storage which allows overwriting of images with new ones. Such archiving systems also allow backing up of files onto second cartridges or laser disks.

e. Colonization of Glass

From the point of both convenience and optical quality, the most practical surface for initial study was the underside of the top coverslip of the ACCMU. Glass also has the advantage that it is biologically inert, so it can be assumed that all nutrition is obtained from the growth medium. Glass is also a representative surface since it forms a significant component, 8.7%, of U.K. refuse.[75] Hence the first studies concentrated on this surface.

In general it was found that the cells of the selected anaerobic associations took at least a week to begin to permanently attach to the underside of the coverslip. During the first week, cells on the surface were loosely attached and, upon subsequent sampling, most were either not found to be present or were not in the same positions as previously. After 2 weeks, cells remained in the same positions and, during this colonization period, were phase dark against a light background.

From week 3, these attached cells became increasingly brighter and appeared to increase in size. This was attributed to the accumulation of external polymers and/or metabolites and transmission of light around and under cells. Cells that had attained a bright appearance were considered to be permanently attached, as they were found to remain in position at subsequent samples. Between cells, surface debris accumulated with time and was thought to be due to free bacterial polymer/metabolites which adhered to the coverslip. It made cells difficult to see and threshold with the image analyser. Reducing the carbon (VFA) concentration in the medium to 5 mM appeared to reduce this problem. Plate 1 (see pages 62 through 64) illustrates a typical colonization sequence over 3 weeks.

Colonization strategies were observed which were comparable with those observed during colonization by aerobic bacteria isolated from natural streams.[70] These included "packing", where cells divide on a surface and the resultant progeny remain attached, resulting in tightly packed colonies, and "spreading", in which the progeny remains attached but spreads outward from the parental cell, forming looser aggregates, often of distinctive shapes, depending on the mode of spread.

The strategy of "shedding",[70] in which progeny does not attach but is dispersed into suspension, was inferred by the fact that some cell types did not form colonies but were homogeneously or randomly distributed. This strategy, and the fourth strategy of "rolling",[70] where progeny is shed by cells rolling across the surface, is more difficult to confirm for anaerobic microorganisms, due to the prolonged observation time demanded by their slow growth rate. It is also possible that some motile cell types shed progeny while in suspension, a hypothesis supported by the observation of motile cells in the medium, clearly in the process of division (Plate 2; see page 64).

Inoculation of the outflow end of the ACCMU chamber was designed to prevent colonization of the uninoculated area by cells carried downstream, so that observed colonization reflected the ability of cells to colonize against the flow of medium, as noted by Korber et al.[76]

Colonies and individual bacteria advanced over the surface of the glass, forming a "colonization front." By noting the position of the colonization front in sequential samples, the rate of colonization was found to be up to 5 mm/d, though 2 to 3 mm/d was most common. Thus, the coverslip was usually colonized to the inflow in 2 to 3 weeks. These rapidly colonizing bacteria predominated in the central region of the chamber, where flow velocity is highest.[64] Lawrence et al.[66] measured flow velocity in a chamber of the same dimensions and found flow velocity within 0.2 μm of the surface to be 0.3% of that in the bulk phase. This boundary layer effect can reduce the nutrient supply obtained from solution near surfaces.[67] Nutrient limitation is likely to have the most effect on those species with the highest growth rate.

In contrast, nonmotile cells and cells of lower colonizing ability were either confined to the inoculum zone or restricted to the sides of the chamber. These included most of the methanogens. After 3 weeks, the older parts of the biofilm, at the outflow, were several micrometers thick in the chamber center and cells could be focused onto approximately 4 to 5 μm deep. However, cells at the sides still formed only a monolayer, possibly due to lower nutrition.

Little change occurred at the coverslip surface, once fully colonized, and so sequential studies were concentrated at the empty areas ahead of the colonization front. As the colonizing cells reached the fields, their modes of colonization could be seen.

E. EXPERIMENTAL STUDIES

1. ACCMU: Identification of Cell Types

Little is known about the species of anaerobic microorganisms that comprise associations in landfill, nor their source of origin. While it may be speculated that the organisms are already present in the refuse as it is deposited, it is also conceivable that organisms living in the deeper, anaerobic layers of the soil may migrate upward into the refuse, once the ground has been sealed off from the atmosphere. It seems likely that both sources contribute to the landfill microbial "cocktail" but evidence for or against any of these possibilities is lacking. Taxonomic studies of refuse components, landfill refuse, and soil anaerobic populations are needed in order to answer these questions. Those studies that have identified populations in refuse have either described total population numbers[77-79] or identified individual species,[80] usually by isolation into monoculture and classic enumeration techniques. Palmisano et al.,[79] for example, determined fermentative bacteria degrading protein or starch in refuse excavated from Fresh Kills and Los Reales landfill sites, and noted that amylolytic bacteria comprised 0.7 to 5.3%, and proteolytic bacteria 0 to 4% of the fermentative population. No cellulolytic organisms were isolated from any samples. However, the authors pointed out that enumeration of bacteria on solid media provides only a conservative estimate of the total number of fermentative bacteria in landfills since only some will be culturable on the medium used. Moreover, the isolation of a viable organism from a landfill site demonstrates only that it survives under landfill conditions and does not imply that the organism is active *in situ*. Such studies also fail to address the nature of microbial interspecies interactions, their influence on degradative processes, and the role surface attachment may play in these interactions. A limited number of studies[81,82] have addressed the effects of immobilization on microbial activity and interactions in anaerobic digesters. Dwyer et al.,[81] for example, studied the effect of immobilization in agar of a methanogenic phenol-degrading association and noted that immobilization increased the apparent K_i from 900 to 1725 mg/l. It is clear, therefore, that we need to identify not only the populations in anaerobic refuse samples but also their attachment and growth on surfaces, and the nature of the interactions on these materials.

The ACCMU system has the potential to show the morphological cell types present in landfill and, by observing changes in species dominance in response to treatments, say something as to the potential physiological roles of the most frequently observed cell types.

To date, approximately 35 different cell types have been recognized by use of the ACCMU. Some are common and were seen in most of the ACCMU studies, while others have only been seen once or twice.

Clumps of spherical sarcina-type cells, which fluoresced blue-green, showed limited colonization in an upstream direction and were found mostly at the outflow end. These are believed to belong to *Methanosarcina*,[83] a nonmotile genus of methanogens. Other similarly fluorescent rod-shaped bacteria were found at higher densities near the sides of the chamber than in the center. Often they were linked together, into filaments, resembling the morphology found in the genus *Methanothrix*.[84] Long, (>100 μm) spiral/undulating, but unidirectional, filaments resembled *Methanospirillum*.[83] Others were short, separate rods, possibly *Methanobacterium*.[83]

Few of the motile cells in suspension were fluorescent. This and the fact that there was little colonization by methanogens outside the inoculum zone, i.e., against the flow, suggests that the methanogens and their progeny were nonmotile. If so, this could place a limitation on the rate at which methanogens can spread throughout a landfill, since distribution would be reliant on passive means of dispersal.

Many methanogens, clearly visible under fluorescence illumination, were obscure or invisible under phase contrast. They were distributed between the phase contrast visible cells and were often obscured by debris. It was in such circumstances that the technique of capturing both phase contrast and fluorescent images of the same field was of particular value as it allowed the position of methanogens to be located on the phase contrast image (Plate 3; see page 65).

Common among the phase contrast visible population were curved and straight rods with a terminal spore, designated LR3 and LR4, respectively. These were possibly sulfate-reducing bacteria (SRB), the curved type resembling species of the genus *Desulfotomaculum*. Type LR3 was distributed randomly and attached to the glass side (Plate 1b; see page 62), suggesting that progeny was shed into suspension. Type LR4 invaded new areas as individual cells, but eventually attached permanently by the spores, which were thus recorded at the center of a colony. This cell type was unlikely to be a sulfate reducer as its abundance decreased markedly when sulfate concentrations were increased from 1 to 5 mM, possibly due to competition for available substrates/nutrients. Where a hexanoate-selected inoculum was used, LR4 was absent from or at low frequency in treatments supplied with butyrate or acetate, but was common to dominant in treatments fed with hexanoate. This indicates an ecological role for LR4 in the degradation of longer-chain fatty acids such as hexanoate.

2. Population Studies in Multi-Stage Continuous Culture Models

Howgrave-Graham et al.[85] used visible and fluorescent microscopy in association with image analysis to quantify planktonic and biofilm samples at different vessel depths within each vessel of a cellobiose-supplemented three-stage anaerobic digester. At imposed dilution rates in vessels A, B, and C of 0.076, 0.030, and 0.020/h, respectively, maintenance of the cellobiose-degraders and sulfate reducers in vessel A was recorded, while 80% of the methanogenic activity was displaced into vessel B. Since only 12 and 8% were located in vessels A and C, respectively, there was significant segregation of these hydrogen sink bacteria. Biofilms were allowed to develop on microscope slides suspended just below each vessel's surface and at its bottom. At the top of vessel B, a biofilm 21 μm thick developed in 13 days; attached colonies of brightly fluorescent short rods and *Methanosarcina*-type cells were interspersed with fluorescent cells, in an attached unilayer of short rods, which comprised 4.5% of the cells (7% of the total cell area). At the bottom of the vessel, where the biofilm was 12 μm thick, 42% of the cell area was fluorescent, due to the presence of both short rods, which occurred singly or in small colonies, and small *Methanosarcina* clumps. Conversely, less than 1% of planktonic cells fluoresced and did so far less brightly than the attached cells.

In vessel C, the biofilm on the slide suspended at the top of the vessel was much thinner (3 μm) than in vessel B, although fluorescent cells comprised 14% of the total cell area. Fluorescent rods and *Methanothrix*-like filaments predominated over *Methanosarcina*-type cells. At the bottom of the vessel, 77% of attached cells were fluorescent, covering 66% of the cell area, confirming the low activity of other groups of bacteria, such as acetogens, in this vessel. Since *Methanosarcina* and, in vessel C, *Methanothrix* comprised a large proportion of the fluorescent population, it was possible that acetoclastic methanogenesis was a major pathway.

In all the vessels, methanogenic activity was primarily located on surfaces or in clumps, emphasizing the potential role of surfaces in landfill sites and their importance in the interspecies interactions which result in the generation and release of methane from refuse polymers.

REFERENCES

1. Parkes, R. J., Methods for enriching, isolating and analysing microbial communities in laboratory systems, in *Microbial Interactions and Communities*, Vol. 1, Bull, A. T. and Slater, J. H., Eds., Academic Press, London, 1982, chap. 3.

2. Campbell, D. J. V., Absorptive capacity of refuse—Harwell research, in *Landfill Leachate Symposium Papers,* Harwell, 1982, paper 3.
3. Terashima, Y., Urabe, S., and Yoshikawa, K., Optimum sampling of municipal solid waste, *Conserv. Recyc.,* 7, 295, 1984.
4. Bull, A. T., Biodegradation: some attitudes and strategies of microorganisms and microbiologists, in *Contemporary Microbial Ecology,* Ellwood, D. C., Hedger, J. N., Latham, M. J., Lynch, J. M., and Slater, J. H., Eds., Academic Press, London, 1980, 107.
5. Slater, J. H. and Hardman, D. J., Microbial ecology in the laboratory, in *Experimental Microbial Ecology,* Burns, R. G. and Slater, J. H., Eds., Blackwell Scientific, Oxford, 1982, chap. 16.
6. Wimpenny, J. W. T., Lovitt, R. W., and Coombs, J. P., Laboratory model systems for the investigation of spatially and temporally organised ecosystems, in *Microbes in their Natural Environment, SGM Symp. 34,* Slater, J. H., Whittenbury, R., and Wimpenny, J. W. T., Eds., Cambridge University Press, Cambridge, 1983, 67.
7. Parkes, R. J. and Senior, E., Multi-stage chemostats and other models for studying anoxic ecosystems, in *Handbook of Laboratory Model Systems for Microbial Ecosystem Research, Vol. 1,* Wimpenny, J. W. T., Ed., CRC Press, Boca Raton, FL, 1988, 51.
8. Bazin, M. J. and Saunders, P. T., Dynamics of nitrification in a continuous flow system, *Soil Biol. Biochem.,* 5, 531, 1973.
9. Blakey, N.C. and Knox, K., The Biodegradation of Phenol under Anaerobic Conditions by Organisms Present in Leachate from Domestic Refuse, WLR Tech. Note Ser. No. 63, Water Research Centre, Stevenage, U.K., 1978.
10. Tittlebaum, M. E., Organic carbon content stabilisation through landfill leachate recirculation, *J. Water Pollut. Control Fed.,* 54, 428, 1982.
11. Robinson, H. D., Barber, C., and Maris, P. J., Generation and treatment of leachate from domestic wastes in landfill, *Water Pollut. Control,* 81, 465, 1982.
12. Shelton, D. and Tiedje, J., General method for determining anaerobic biodegradation potential, *Appl. Environ. Microbiol.,* 47, 850, 1984.
13. Bogner, J. E., Controlled study of landfill biodegradation rates using modified BMP assays, *Waste Manage. Res.,* 8, 329, 1990.
14. Watson-Craik, I. A., Sinclair, K. J., James, A. G., Sulisti, and Senior, E., Studies of the refuse methanogenic fermentation by use of laboratory systems, *Water Sci. Technol.,* 27, 15, 1993.
15. Warner, J. S., Hidy, B. J., Jungclaus, G. A., McKown, M. M., Miller, M. P., and Riggin, R. M., Development of a method for determining the leachability of organic compounds from solid wastes, in *Hazardous Solid Waste Testing: First Conference,* ASTM Spec. Tech. Publ. 760, Conway, R. A. and Malloy, B. C., Eds., American Society for Testing and Materials, Philadelphia, 1981, 40.
16. Newton, J. R., Pilot-Scale Studies on Leaching from Landfills. III. Leaching of Hazardous Wastes, WLR Tech. Note Ser. No. 51, Department of the Environment, London, 1977.
17. Josephson, J., Immobilisation and leachability of hazardous wastes, *Environ. Sci. Technol.,* 16, 219A, 1982.

18. Jackson, D. R., Garrett, B. C., and Bishop, T. A., Comparison of batch and column methods for assessing leachability of hazardous waste, *Environ. Sci. Technol.*, 18, 668, 1984.

19. Barlaz, M. A., Ham, R. K., and Schaefer, D. M., Microbial, chemical and methane production characteristics of anaerobically decomposed refuse with and without leachate recycling, *Waste Manage. Res.*, 10, 257, 1992.

20. Watson-Craik, I. A. and Senior, E., Landfill co-disposal: hydraulic loading rate considerations, *J. Chem. Technol. Biotechnol.*, 45, 203, 1989.

21. Watson-Craik, I. A. and Senior, E., Treatment of phenolic wastewaters by co-disposal with refuse, *Water Res.*, 23, 1293, 1989.

22. Watson-Craik, I. A. and Senior, E., Landfill co-disposal of phenol-bearing wastewaters: organic load considerations, *J. Chem. Technol. Biotechnol.*, 47, 219, 1990.

23. Cooperative Programme of Research on the Behaviour of Hazardous Wastes in Landfill Sites, Final Report, Department of the Environment, Her Majesty's Stationery Office, London, 1978.

24. Flyvbjerg, J., Arvin, E., Jensen, B. K., and Olsen, S. L., Microbial degradation of phenols and aromatic hydrocarbons in creosote-contaminated groundwater under nitrate-reducing condition, *J. Contam. Hydrol.*, 12, 133, 1993.

25. Godsy, E. M., Goerlitz, D. F., and Erlich, G. G., Methanogenesis of phenolic compounds by a bacterial consortium from a contaminated aquifer in St. Louis Park, Minnesota, *Bull. Environ. Contam. Toxicol.*, 30, 261, 1983.

26. Fedorak, P. M. and Hrudey, S. E., Anaerobic treatment of phenolic coal conversion wastewater in semi-continuous cultures, *Water Res.*, 20, 113, 1986.

27. Wang, Y.-T., Suidan, M. T., Pfeffer, J. F., and Najam, I., The effect of concentration of phenols on their batch methanogenesis, *Biotechnol. Bioeng.*, 33, 1353, 1989.

28. Gallert, C. and Winter, J., Uptake of phenol by the phenolmetabolizing bacteria of a stable, strictly anaerobic consortium, *Appl. Microbiol. Biotechnol.*, 39, 627, 1993.

29. Balba, M. T. M. and Evans, W. C., The methanogenic biodegradation of catechol by a methanogenic consortium: evidence for the production of phenol through *cis*-benzenediol, *Biochem. Soc. Trans.*, 8, 452, 1980.

30. Kaiser, J. P. and Hanselmann, K. W., Aromatic chemicals through anaerobic microbial conversion to lignin monomers, *Experientia*, 38, 167, 1982.

31. Balba, M. T. M. and Evans, W. C., Methanogenic fermentations of the naturally occurring aromatic amino acids by a microbial consortium, *Biochem. Soc. Trans.*, 8, 1979, 1980.

32. Bakker, G., Anaerobic degradation of aromatic compounds in the presence of nitrate, *FEMS Lett.*, 1, 103, 1977.

33. Balba, M. T. M., Clarke, N. A., and Evans, W. C., The methanogenic fermentation of plant phenolics, *Biochem. Soc. Trans.*, 7, 1115, 1979.

34. Fedorak, P. M. and Hrudey, S. E., The effects of phenol and some alkyl phenolics on batch anaerobic methanogenesis, *Water Res.*, 18, 361, 1984.

35. Rees, J. F. and King, J. W., The dynamic of anaerobic phenol degradation in Lower Greensand, *J. Chem. Technol. Biotechnol.*, 31, 306, 1981.

36. Cheyney, A. C., Experience with the co-disposal of hazardous wastes with domestic refuse, *Chem. Ind.,* 17, 609, 1984.

37. Robinson, H. D. and Maris, P. J., Leachate from Domestic Wastes: Generation, Composition and Treatment; A Review, Tech. Rep. TR108, Water Research Centre, Stevenage, U.K., 1979.

38. Pearson, F., Shiun-Chung, C., and Gautier, M., Toxic inhibition of anaerobic degradation, *J. Water Pollut. Control Fed.,* 52, 472, 1980.

39. Gourdon, R., Comel, C., Vermande, P., and Veron, J., *n*-Butyrate catabolism in the treatment of a semi-synthetic landfill leachate on anaerobic filter under sequential conditions — methane production by β-oxidation during continuous feeding and by decarboxylation during sequential feeding, *Biomass,* 15, 11, 1988.

40. Stanforth, R., Ham, R., Anderson, M., and Stegman, R., Development of a synthetic municipal leachate, *J. Water Pollut. Control Fed.,* 51, 1965, 1979.

41. Fungaroli, A. A. and Steiner, R. L., Laboratory study of the behaviour of a sanitary landfill, *J. Water Pollut. Control Fed.,* 43, 252, 1971.

42. Boreham, D., Bromley, J., Parker, A., and Wright, S. J., Unsaturated and Saturated Laboratory Column Experiments, WLR Tech. Note Ser. No. 39/41, DOE, London, 1976.

43. Newton, J. R., Pilot-scale studies on the leaching of industrial wastes in simulated landfills, *Water Pollut. Control,* 76, 468, 1977.

44. Robinson, H. D. and Maris, P. J., The treatment of leachates from domestic wastes in landfill sites, *J. Water Pollut. Control Fed.,* 57, 30, 1985.

45. Bazin, M. J., Saunders, P. T., and Prosser, J. I., Models of microbial interactions in the soil, *Crit. Rev. Microbiol.,* 4, 463, 1976.

46. Macura, J. and Malik, I., Continuous flow method for the study of microbiological processes in soil samples, *Nature,* 182, 1796, 1958.

47. Girard, P., Scharer, J. M., and Moo-Young, M., Two-stage anaerobic digestion for the treatment of cellulosic wastes, *Chem. Eng. J.,* 33, B1, 1986.

48. Gijzen, H. J., Zwart, K. B., Verhagen, F. J., and Vogels, G. D., High-rate two-phase process for the anaerobic degradation of cellulose, employing rumen microorganisms for an efficient acidogenesis, *Biotechnol. Bioeng.,* 31, 418, 1988.

49. Kisaalita, W. S., Lo, K. V., and Pinder, K. L., Influence of dilution rate on the acidogenic phase products distribution during two-phase lactose anaerobiosis, *Biotechnol. Bioeng.,* 34, 1235, 1989.

50. Coutts, D. A. P., Senior, E., and Balba, M. T. M., Multi-stage chemostat investigation of interspecies interactions in a hexanoate-catabolising microbial association isolated from anoxic landfill, *J. Appl. Bacteriol.,* 623, 251, 1987.

51. Harmsen, J., Identification of organic compounds in leachate from a waste tip, *Water Res.,* 17, 699, 1983.

52. Watson-Craik, I. A., James, A. G., and Senior, E., Use of multi-stage continuous culture systems to investigate the effects of temperature on the methanogenic fermentation of cellulose-degradation intermediates, in *Proc. 7th Int. Symposium Anaerobic Digestion,* International Association on Water Quality, Cape Town, 1994, 2.

53. Rees, J. F., The fate of carbon compounds in the landfill disposal of organic matter, *J. Chem. Technol. Biotechnol.,* 30, 161, 1980.
54. Rees, J. F., Major factors affecting methane production in landfills, in *Landfill Gas Symposium Papers,* U.K. Atomic Energy Authority, Harwell, 1981, paper 3.
55. Watson-Craik, I. A., Landfill as an Anaerobic Filter for the Co-Disposal of Phenolic Wastewaters, Ph.D. Thesis, University of Strathclyde, Glasgow, 1987.
56. Watson-Craik, I. A. and Senior, E., Effect of phenol wastewater co-disposal on the attenuation of the refuse leachate molecule hexanoic acid, *Lett. Appl. Microbiol.,* 9, 227, 1989.
57. Perfilev, B. V. and Gabe, D. R., *Capillary Methods of Investigating Microorganisms,* Shewan, J. M., Transl., University of Toronto Press, Toronto, 1969.
58. Quesnel, L. B., Methods of microculture, in *Methods in Microbiology,* Vol. 1, Norris, J. R. and Ribbons, D. W., Eds., Academic Press, London, 1969, 365.
59. Harris, N. K. and Powell, E. O., A culture chamber for the microscopical study of living bacteria with some observations on the spore-bearing aerobes, *J. Royal Microsc. Soc.,* 70, 407, 1951.
60. Duxbury, T., A microperfusion chamber for studying the growth of bacterial cells, *J. Appl. Bacteriol.,* 43, 247, 1977.
61. Berg, H. E. and Block, S. M., A miniature flow cell designed for rapid exchange of media under high-powered microscope objectives, *J. Gen. Microbiol.,* 130, 2915, 1984.
62. Sjollema, J., Busscher, H. J., and Weerkamp, A. H., Real-time enumeration of adhering microorganisms in a parallel plateflow cell using automated image analysis, *J. Microbiol. Methods,* 9, 73, 1989.
63. Rutter, P. and Leech, R., The deposition of Streptococcus sanguis NCTC 7868 from a flowing suspension, *J. Gen. Microbiol.,* 120, 301, 1980.
64. Caldwell, D. E. and Lawrence, J. R., Study of attached cells in continuous-flow slide culture, in *CRC Handbook of Laboratory Model Systems for Microbial Ecosystems,* Vol. 1, Wimpenny, J. W. T., Ed., CRC Press, Boca Raton, FL, 1988, 117.
65. Caldwell, D. E. and Lawrence, J. R., Growth kinetics of *Pseudomonas fluorescens* microcolonies within the hydrodynamic boundary layers of surface microenvironments, *Microb. Ecol.,* 12, 299, 1986.
66. Lawrence, J. R., Delaquis, P. J., Korber, D. R., and Caldwell, D. E., Behaviour of *Pseudomonas fluorescens* within the hydrodynamic boundary layers of surface microenvironments, *Microb. Ecol.,* 14, 1, 1987.
67. Lawrence, J. R., Korber, D. R., and Caldwell, D. E., Computer-enhanced darkfield microscopy for the quantitative analysis of bacterial growth and behaviour on surfaces, *J. Microbiol. Methods,* 10, 123, 1989.
68. Caldwell, D. E. and Germida, J. J., Evaluation of difference imagery for visualising and quantifying microbial growth, *Can. J. Microbiol.,* 31, 35, 1985.
69. Caldwell, D. E., New developments in computer-enhanced microscopy, *J. Microbiol. Methods,* 4, 117, 1985.

70. Lawrence, J. R. and Caldwell, D. E., Behaviour of bacterial stream populations within the hydrodynamic boundary layers of surface microenvironments, *Microb. Ecol.,* 14, 15, 1987.

71. Balba, M. T. M., Part 1: The Methanogenic Fermentation of Aromatic Substrates. Part 2: The Origin of Hexahydrohippurate in the Urine of Herbivores, Ph.D. Thesis, University of Wales, U.K., 1978.

72. Coutts, D. A., Balba, M. T. M., and Senior, E., Acetogenesis and acetotrophy in a hexanoate-catabolising microbial association isolated from anoxic landfill, *J. Appl. Bacteriol.,* 63, 343, 1987.

73. Mink, R. W. and Dugan, P. R., Tentative identification of methanogenic bacteria by fluorescence microscopy, *Appl. Environ. Microbiol.,* 33, 713, 1977.

74. Kuhn, D. A. and Starr, M. P., Effects of microscope illumination on bacterial development, *Arch. Microbiol.,* 74, 292, 1970.

75. Wigfull, S. and Gregory, R. G., Prediction of the effect of recycling initiatives on waste composition and the national landfill gas resource, in Landfill Microbiology: R & D Workshop II, Evans, S. A. and Lawson, P. S., Eds., Harwell Laboratories, Harwell, U.K., 991, 35.

76. Korber, D. R., Lawrence, J. R., Sutton, B., and Caldwell, D. E., Effect of laminar flow velocity on the kinetics of surface recolonization by Mot$^+$ and Mot$^-$ *Pseudomonas fluorescens, Microb. Ecol.,* 18, 1, 1989.

77. Pahren, H. R., Microorganisms in municipal solid waste and public health implications, *CRC Crit. Rev. Environ. Control,* 17, 187, 1987.

78. Barlaz, M. A., Schaefer, D. M., and Ham, R. K., Bacterial population development and chemical characteristics of refuse decomposition in a simulated sanitary landfill, *Appl. Environ. Microbiol.,* 55, 55, 1989.

79. Palmisano, A. C., Maruscik, D. A., and Schwab, B. S., Enumeration of fermentative and hydrolytic micro-organisms from three sanitary landfills, *J. Gen. Microbiol.,* 139, 387, 1993.

80. Fielding, E. R., Archer, D. B., Conway de Macario, E., and Macario, A., Isolation and characterization of methanogenic bacteria from landfills, *Appl. Environ. Microbiol.,* 54, 835, 1988.

81. Dwyer, D. F., Krumme, M. L., Boyd, S. A., and Tiedje, J. M., Kinetics of phenol biodegradation by an immobilised methanogenic consortium, *Appl. Environ. Microbiol.,* 52, 345, 1986.

82. Isa, Z., Grusenmeyer, S., and Verstraete, W., Sulfate reduction relative to methane production in high-rate anaerobic digestion: microbiological aspects, *Appl. Environ. Microbiol.,* 51, 580, 1986.

83. Zeikus, J. G. and Bowen, V. G., Comparative ultrastructure of methanogenic bacteria, *Can. J. Microbiol.,* 21, 121, 1975.

84. Kobayashi, H. A., Conway de Macario, E., Williams, R. S., and Macario, A. J. L., Direct characterization of methanogens in two high-rate anaerobic biological reactors, *Appl. Environ. Microbiol.,* 54, 693, 1988.

85. Howgrave-Graham, A. G., Jones, L. R., James, A. G., Terry, S. J., Senior, E., and Watson-Craik, I. A., Microbial distribution throughout a cellobiose-supplemented three-stage laboratory-scale anaerobic digester, *J. Chem. Technol. Biotechnol.,* 59, 127, 1994.

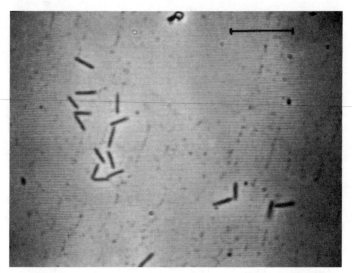

a

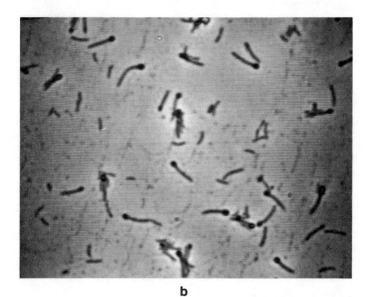

b

Plate 1   Colonization of glass over a 3 week period: (a and b), from day 3 to day 7, cells are not permanently attached, so vary in position and number; (c), day 11, the start of cell attachment and debris accumulation; (d and e), days 14 and 21, permanent cell attachment with cells becoming brighter. Bar denotes 10 μm.

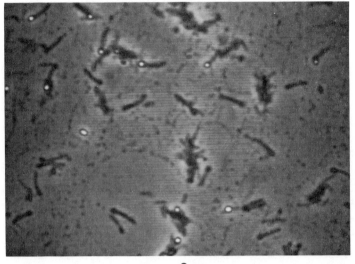

c

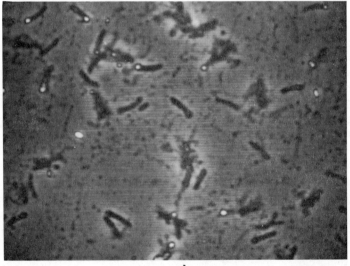

d

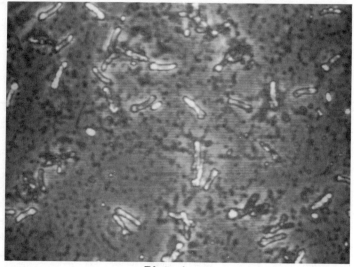

Plate 1e

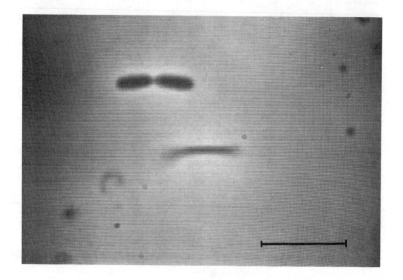

Plate 2   A pair of motile cells in suspension. Bar denotes 5 μm.

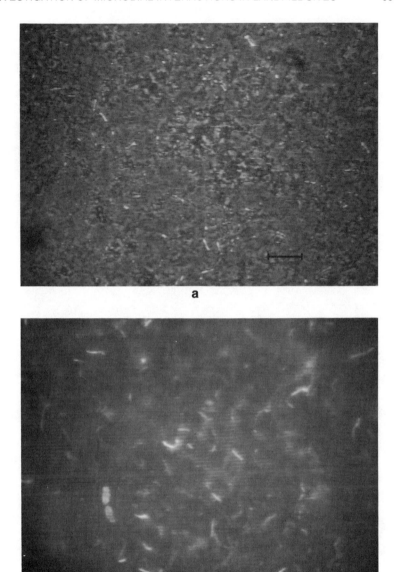

Plate 3   (a) Phase contrast and (b) F_{420} fluorescence images of the same field can indicate the location of methanogenesis in a complex biofilm. Bar denotes 10 μm.

CHAPTER **3**

Mathematical Modeling of the Methanogenic Ecosystem

Alan Young

CONTENTS

0-87371-968-9/95/$0.00+$.50
© 1995 by CRC Press, Inc.

I. INTRODUCTION

A mathematical model may be thought of as a way of combining a number of subsystems, each of which is understood independently, into a whole assembly which is not otherwise predictable. Landfills are extremely complex environments in which bacteria are both controlled by and exert control over the local ecosphere, and simple laboratory experiments have not been able to replicate the full extent of these interactions. Hence, numerous attempts have been made to convert existing knowledge into mathematical formulations and then computer programs.

The steps involved in constructing a computer model are as follows:

1. Decide what information is being sought.
2. Identify the external controllable influences.
3. Identify the main internal factors involved in the system.
4. Describe the way in which each of the internal factors responds to changes in each other and the external influences.
5. Translate the mathematical description into computer code.
6. Select initial conditions and run program.

This chapter illustrates the process (including its benefits and pitfalls) using part of a model produced at Oxford University for the U.K. Department of the Environment. The full system is described in CWM039/92, but herein the focus is on the aspect most closely concerned with the microbiology. In this part of the project, the early stages were roughly

1. What determines whether methanogenesis begins at a site, how robust is it, and how long does it take to become established? To what extent can site operators control the development of a site?
2. Waste composition, size, and shape of site, water input, pretreatment of waste.
3. Temperature, moisture level, pH, E_h, bacterial population, solute concentrations.
4. A mixture of laboratory and field data was used.
5. The specification was made in terms of a set of differential equations and programmed in 'C' and Fortran.
6. A large range of initial conditions were run, with the aim of producing a set of generic guidelines for landfill behavior.

A significant by-product of the modeling process (at step 4) was the identification of areas in which fundamental data are lacking, and of the need to perform experiments before further progress can be made. One advantage of modeling over experimentation is that there are no uncontrollable factors and full information is available when trying to understand simulation output. These results vindicated the initial contention that all the mechanisms involved in regulating landfill degradation interact and cannot ultimately be considered in isolation from each other. Hence, one of the main results for the project was to advocate a more holistic approach to landfill management, rather than concentrating on one or two supposedly key parameters as most previous studies had done.

II. RATIONALE

After the first year or so of landfill decomposition the major gaseous products are normally carbon dioxide and methane, and the reactions producing them continue at a gradually decreasing rate for at least 15 and often upwards of 50 years. It is generally accepted that the majority of the degradable waste will decompose during this latter phase. The onset of large-scale methane production is thus of great importance when managing a site, both in terms of its environmental impact and the commercial opportunities for its exploitation. Before this occurs the gas vented is largely carbon dioxide with an admixture of hydrogen, and the leachate is highly acidic. When the methane-producing phase is entered, hydrogen ceases to be found in the gas, and the leachate becomes much less acidic (and consequently less harmful in the event it escapes). The methanogenic bacteria responsible for this are regarded as being among the least robust species found within a domestic landfill, and sometimes fail entirely to develop.

Hence, the primary aim of this section of the Oxford model was to predict the conditions under which methanogenesis is able to become established, to estimate the time required for this to happen, and to investigate methods of influencing this period.

The difficulties in gathering detailed information on the composition of wastes deposited at a particular site make it impossible to produce specific predictions of the site's future behavior—in addition to which the computer time required would be prohibitive. Instead, effort was concentrated on producing a generic model to identify the primary *in situ* interactions and any feasible methods of controlling them.

III. METHODOLOGY

The actual reactions occurring within a landfill are too numerous and complex to model precisely. To make the problem manageable, the substances involved must be grouped into a number of generic categories and the system phrased in terms of these. It is thought that the most important metabolic pathway in refuse decomposition is fermentation of the primary substrates (mainly paper and vegetative matter) to sugars and alcohols followed by their conversion to carboxylic acids, which may then break down

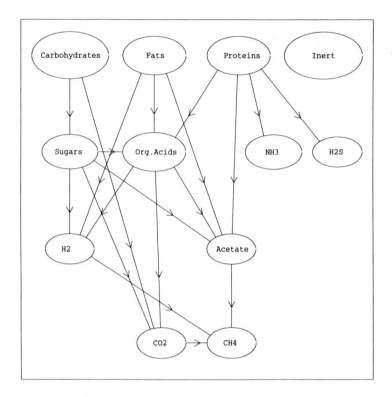

Figure 1   Modeled biochemical processes.

to produce methane. Figure 1 illustrates the divisions that were chosen as being the best simple representation of this system, with the arrows indicating the overall direction of mass conversion. Data derived from samples of landfill waste and leachate were used to estimate an impirical formula for each of the composite substances, and the proportions in which they transformed to one another were based on simple chemical relations (see following).

The multiphase nature of the landfill environment and the influence of biological factors upon degradation introduce significant complications. The former introduces spatial variability into the problem and the latter makes application of standard chemical kinetic theory highly unreliable. The action of the bacteria within a site is a primary mechanism regulating their temperature, chemical composition, and physical structure (i.e., the saturation, permeability, and density) and these factors, in turn, determine the viability of the many microbial species present.

The initial phases of decomposition and acidogenesis are carried out by a large, ill-defined group of species whose diversity makes the processes fairly robust so that they are probably found in all sites. By contrast, Archer and Fielding[1] claimed that there are only a limited number of common methanogenic varieties (the archaebacteria), and that these are more sensitive to adverse conditions, being significantly slower growing than most of the other types and sometimes failing entirely to survive. Considering their different behaviors, and that the level of organics potentially able to escape from a site is crucially dependent on the amount of acids consumed by methane production, it was decided to model the populations of methanogenic bacteria explicitly and treat the earlier processes as being dependent purely on the instantaneous state of substrate/nutrient availability (thus assuming that the size of their populations will adapt very rapidly to the local steady-state level).

IV. IDEALIZED REACTIONS

The reactions relating the transformation of mass between the substances were taken to be as below. No assertion is made that these are the precise reactions occurring, only that they provide a plausible way to relate the mass transference between compounds.

Hydrolysis and fermentation of carbohydrate, $n = 1, \ldots$

$$(CH_2O)_{3n} + 2H_2O \rightarrow 2H(CH_2)_nOH + nCO_2 + nH_2O \tag{1}$$

Degradation* of proteins, the intermediate stage of deamination being ignored

$$C_{46}H_{77}O_{17}N_{12}S + \{19.95\}H_2O \rightarrow \{0.421\}C_{69}H_{138}O_{32} \tag{2}$$
$$+ \{5.19\}CH_3COOH + \{6.545\}CO_2 + 12NH_3 + H_2S$$

Degradation of fats, the intermediate stage of hydrolysis being ignored

$$C_{55}H_{104}O_6 + \{27.27\}H_2O \rightarrow \{0.587\}C_{69}H_{138}O_{32} \tag{3}$$
$$+ \{7.23\}CH_3COOH + \{24.27\}H_2$$

Acidogenesis from glucose

$$C_6H_{12}O_6 + (6 - 2n)H_2O \rightarrow H(CH_2)_n COOH + (11 - 3n)H_2 + (5 - n)CO_2 \tag{4}$$

Acidogenesis from the alcohols

$$CH_3(CH_2)_nOH + H_2O \rightarrow H(CH_2)_nCOOH + 2H_2 \tag{5}$$

---

* The brackets { } indicate approximate values. Since the reactions are not precise they would require large coefficients to balance them exactly.

Breakdown of carboxylic acids, $n = 2, 3, 4, 5$

$$H(CH_2)_nCOOH + 2(n - 1)H_2O \rightarrow CH_3COOH + 3(n - 1)H_2 + (n - 1)CO_2 \quad (6)$$

Breakdown of acetic acid

$$CH_3COOH \rightarrow CH_4 + CO_2 \quad (7)$$

Consumption of hydrogen

$$4H_2 + CO_2 \rightarrow CH_4 + 2H_2O \quad (8)$$

V. TRANSPORT EQUATIONS

Several of the major reactive components within a landfill can exist in both gaseous and dissolved states. To allow for these transferences the interior was assumed to be divided into three overlapping zones: water, gas, and solid waste—occupying volume fractions θ, ϕ, and $1 - (\theta + \phi)$, respectively.

The combined set of transport equations for a single chemical species are thus

$$\frac{\partial}{\partial t}(M_\theta) + \underline{v}. \nabla (M_\theta/\theta) = \text{div}(\underline{\underline{E}}\nabla M_\theta) + q - Q \quad (9)$$

$$\phi\frac{\partial}{\partial t}(M\phi) = \text{div}(M_\phi\underline{\underline{K}} \cdot \nabla P) + \phi Q \quad (10)$$

where $M\phi$, M_ϕ are the masses of compound present in each state per unit volume of the landfill; $\underline{v}$ is the imposed Darcy water velocity; $\underline{\underline{E}}$ the combined diffusion and dispersion tensor; $\underline{\underline{K}}$ the gas permeability tensor; and P the gas pressure within the voids. q represents the net rate of formation of the compound within the liquid medium; and Q represents the evolution of dissolved gases from the leachate. For a solute such as acetic acid, with no gaseous form, the Q term vanishes and Equation 10 is irrelevant.

Even if q and Q were known precisely, the pair (Equations 9 and 10) is difficult to solve for a single solute, and since several interacting compounds are present the extended set of equations is insoluble by current analytical methods. In view of this complexity and the difficulty of interpreting the results of a full numerical solution, it is necessary to verify the biochemical relations before introducing spatial variability (the project had not reached this latter stage by its termination date in 1992). To do this, the landfill is assumed to be a "well-mixed" reactor, so that the equations reduce to

$$\frac{d}{dt}(M) = q - Q + \Omega \tag{11}$$

$$\frac{d}{dt}(\theta\rho) = Q - V \tag{12}$$

where M is the total dissolved mass; ρ the density of the gas phase; V represents the rate of gas venting when the system exceeds atmospheric pressure; and Ω is the flux of solute in and out of the landfill due to a prescribed water flow (V and Ω thus compensate for the boundary conditions lost during the simplification).

The dissolved and gaseous components of a solute are, provided it does not react chemically with water, related by Henry's law. No suitable data were available with which to predict the rate of gas evolution/solution when the system is not in equilibrium, and this process is also likely to be greatly influenced by local factors such as the area of the liquid-gas interface, so a sensitivity analysis of the effect of different forms for Q was carried out by performing a series of numerical simulations.

VI. RATE-CONTROLLING FACTORS

In order to model the biochemical degradation it is necessary to derive expressions for the rates at which each of the major individual processes are taking place. Most of the important reactions are performed by bacteria within the site and this prevents standard chemical results being applied. For example, above 60°C the mesophilic methanogens seem to become extinct (or are rendered inactive) and the cleavage of acetate via Reaction 7 usually ceases. A survey of the literature indicated that the primary influences on anaerobic bacterial activity are temperature, pH, E_h, substrate availability, moisture level, oxygen content, trace nutrients, and the presence of toxins.

The possible presence of energetically advantaged sulfate-reducing bacteria (which may out-compete the methanogens for hydrogen) was allowed for by assuming that any sulfate produced was immediately metabolized to sulfide. Aerobic methane-consuming species were neglected (since they require oxygen and are, thus, normally found only in a shallow surface zone) since, although they may alter the apparent temperature and rate/concentration of gas production as measured at the surface, it is assumed that they do not appreciably influence the real internal generation rate (with which this part of the model is concerned).

As degradation proceeds the solid primary substrates will be broken down and their volume and strength decrease. Some settlement will occur, but at a lesser rate than the waste decomposition suggests, and to simplify the model it was assumed that this settlement can be ignored during the first year. The solid volume occupied was thus taken to be proportional to the mass remaining as a fraction of that originally present. As new void space is created the evolved gas is easily able to fill it, so that no allowance is necessary for air being drawn in.

A. PRIMARY DECOMPOSITION

The initial degradation by fermentation, hydrolysis, and glycolysis can take place via a large number of metabolic pathways depending on the exact chemical composition of the substrate, and little is known of them individually. Instead of laboratory data it is necessary to rely on more empirical information from sewage digester and landfill measurements, and it was decided to extend the approach of Hoeks[2] as the best way to make use of this. The waste is categorized as a mixture of four principal components: carbohydrates, proteins, fats, and biologically inert materials (e.g., masses $G_1, \ldots, G_4$), and the first three of these are further divided into slow, medium, and readily degradable groups. This division of waste into ten groups allows considerable flexibility when specifying the initial composition of the simulated landfill.

To introduce feedback into the model the decomposition rates of each subcategory were taken to be functions of temperature, pH, and the local water saturation level (though no allowance could be made for the increased mobility of nutrients in wetter regions).

Using mainly digester data from Coleman et al.[3] and Huang et al.,[4] together with general microbiological principles (for the behavior above 55°C), the normalized rate of reaction with temperature was assumed to be as in Figure 2.

Quantitative information on the pH behavior of the main landfill bacteria under field conditions is sparse outside of the range 5.5 < pH < 7.5. The "bell-shaped" plot of Figure 3 represents an empirical estimate of this dependency.

Most of the degradative bacteria require an aqueous environment, and it seems reasonable that the area of waste open to decomposition is related to the degree of saturation. In the absence of quantitative information, this dependency was chosen to be directly proportional to the fraction of water present compared to its maximum saturated value. Particle size is also important, but this is not quantified for unprocessed waste and thus, could not be included as a separate factor.

Combining these factors, it was proposed that the mass of the ith type of waste, of decay category j, obeys the first order equation:

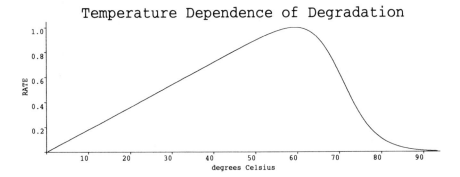

Figure 2 Normalized modeled rate of degradation vs. temperature.

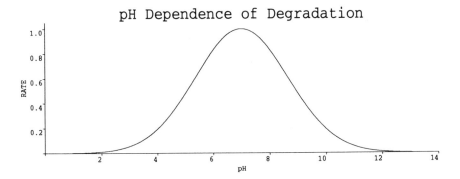

Figure 3 Normalized modeled rate of degradation vs. pH.

$$\frac{d}{dt}(G_{ij}) = -v_j\xi(T, \text{pH}, \phi)G_{ij} \tag{13}$$

with the v_j being empirical rate constants (in s^{-1}) and

$$\xi(T, pH, \beta) = \frac{T\phi\exp\left[-\{\text{pH} - 7\}^2 \ln(4/3)\right]}{\phi_{max}\{1 + \exp(T/4 - 18)\}} \tag{14}$$

with T being the temperature in Celsius. Choosing i

$$v_1 = 6.29 \times 10^{-10} \tag{15}$$

$$v_2 = 1.26 \times 10^{-10} \tag{16}$$

$$v_3 = 2.10 \times 10^{-11}{}^{\circ}C^{-1} s^{-1} \tag{17}$$

means that under constant conditions of $T = 35°C$, pH $= 7$, $\phi = \phi_{max}$ the three types of waste would have half-lives of 1, 5, and 30 years, respectively. Findikakis and Leckie[5] suggested that fast, medium, and slowly degradable wastes might be present in the relative proportions of 15, 55, and 30% in fresh waste—through their definition of the categories varies between sites.

Note that this simplification reduces the models predictive power regarding the longer term course (greater than a 5 years) of degradation but is adequate for the initial phase which was the desired goal.

B. SECONDARY DEGRADATION

The carbohydrate materials are modeled as breaking down into soluble sugars and alcohols (in the ratio of 2:1 by mass) and thence to carboxylic acids, whereas this intermediate stage is neglected for the proteinaceous and fatty materials. There is no direct evidence for the existence of free alcohols in landfill leachate but this proved impossible to obtain as they are not normally tested for.

The levels of sugars observed are small compared to the acid concentrations so that the associated bacterial populations seem to respond rapidly to changes in substrate availability. This phase of acidogenesis was treated as a simple reaction, with the rate of utilization proportional to the combined glucose/alcohol concentration Z_b thus

$$\frac{dZ_b}{dt} = -\lambda_b Z_b \tag{18}$$

taking $\lambda_b = 8 \times 10^{-6}/s$ so that in the absence of further production it has a "half-life" of 1 d.

It is useful to distinguish acetic acid from the longer-chain carboxylic acids (represented by a composite formula) since the latter are not observed to form a direct substrate for methanogenesis (substances such as formate have been assumed to be insignificant). The primary substrates and glucose/alcohol solutes are modeled as decomposing directly into acetic acid and the composite acid in the ratio of 1:2 by mass, and the succeeding process, whereby the long-chain acids break down into acetic acid, is treated separately. It is thought that this latter acetogenesis is inhibited greatly by the presence of dissolved hydrogen, and that this may be the dominating factor determining its velocity. Numerical data concerning this effect are scarce, though Archer and Harris[6] asserted that a "well-established" methanogenic system usually has around $0.2 \mu g/dm^3$ dissolved hydrogen, whereas the peak partial pressure observed is about 30 kPa which corresponds to 0.3 mg/dm^3 at 30°C. This relation is especially difficult to analyze due to the formation of syntrophic flocs of methanogens and acidogens, which means that the concentration levels perceived by the bacteria may be quite different from the average measured solution values.

In the absence of quantitative information, the free energy equation was used as a qualitative guideline, noting that 1 mol of the composite acid decomposes to produce 4.63 mols of hydrogen, and the level of carbon dioxide is relatively constant. The relation assumed was

$$\frac{dX_A}{dt} = -\lambda_A X_A \max\left\{0, \ln\left(\frac{Z_A}{Z_a Z_h^{4.63}}\right)\right\} \qquad (19)$$

where Z_h, Z_a, and Z_A are the dissolved concentrations of hydrogen, acetic and composite acid in mg/dm³, X_A being the dissolved mass of composite acid in mg/dm³ of the total landfill volume. Choosing $\lambda_A = 2.78 \times 10^{-8}/s^1$ means that, when $Z_a = Z_A$ and $Z_h = 1.35 \times 10^{-4}$mg/dm³, the composite acid will have a half-life of 1 week in the absence of further production.

C. METHANOGENESIS

The methanogenic bacteria were modeled as two distinct species, labeled "a" (acetoclastic) and "h" (hydrogen consuming) which gain their energy from Reactions 7 and 8, respectively, though some species have been observed to perform both. The masses of these species are denoted by X_a and X_h mg/dm³. These species are not strongly motile (some researchers suggest they are static) and may be modeled as a simple solute using Equation 9, so that they obey Equation 11 in the simplified case. Pirt[7] asserted that their reproductive behavior may be described by

$$\frac{dX_a}{dt} = (Y_a E_a - k_{ad})X_a \qquad (20)$$

$$\frac{dX_h}{dt} = (Y_h E_h - k_{hd})X_h \qquad (21)$$

where Y is the growth yield coefficient (the mass of bacteria produced per mass of substrate consumed); E the rate of substrate utilization; and k_d the endogenous death rate which allows for such factors as maintenance energy. The magnitudes of Y and k_d vary with the environment but their form seems unknown, so they were assigned the constant values $Y_h = 0.3$, $k_{hd} = 0.022/d^1 = k_{ad}$ and $Y_a = 0.048$ using the data of Lawrence and McCarty.[8] The rate of substrate utilization was assumed to follow Monod's law (though Farquhar and Rovers[9] qualified this by indicating that it may be inaccurate for acetate concentrations above 3000 mg/dm³), with the half-velocity and maximum rate governed by nonsubstrate factors.

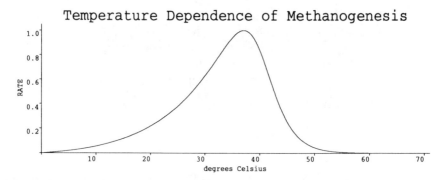

Figure 4   Normalized modeled rate of methanogenesis vs. temperature.

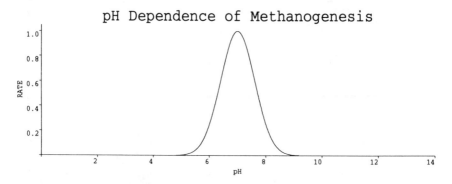

Figure 5   Normalized modeled rate of methanogenesis vs. pH.

Information is lacking regarding the functional form of E, but Figures 4 and 5 show the observed behavior of methane production with temperature and pH, and it is assumed that this followed the activity of the methanogens and functions were fit to the curves. A special study was commissioned by Strathclyde University[10] to investigate the effects of pH and temperature on methanogens. The temperature results echoed the general literature, while the pH experiments indicated that the bacteria are significantly more tolerant of acidic conditions than is normally assumed. The data available provide no way to differentiate between the behaviors of the various species of methanogens so the temperature and pH dependency were taken to be the same for both of the types. It was impossible to assess the relative importance of direct substrate inhibition vs. indirect pH-inhibition effects in the acetoclastic reactions.

The half velocities K_i in Monod's law were taken from Zinder[11] as K_h = 0.005 mg/dm^3 H_2 (it being observed that the modeled carbon dioxide was always present in excess), and for the acetate consumers a temperature dependency was included using data from Lawrence and McCarty[8] to give

$$K_a = \begin{cases} 869 \text{ mg/dm}^3, & T \leq 25°C \\ \text{linear interpolation,} & 25 < T < 35 \\ 172 \text{ mg/dm}^3, & T \geq 35°C \end{cases} \qquad (22)$$

where the additional value $K_a = 333$ mg/dm^3 at 30°C was also used for the intermediate interpolation. The overall form of E used was therefore

$$E_i = E_{max,i} \frac{Z_i \exp[-(pH - 7)^2 \ln(4)](e^{-aT} - e^{-bT})}{(K_i + Z_i)(1 + e^{-c(T - d)})} \qquad (23)$$

where Z_i is the dissolved concentration of substrate for the i bacteria; and K_i is the Monod half-velocity. The coefficients a, b, c, and d are used to fit the curve of Figure 4, and have the values 0.08, 0.05, 0.45, and 40, respectively. The absolute maximum rates of consumption under optimum conditions were taken to be $E_{max.m} = 1.86/d$ and $E_{max,M} = 8/d$.

Figure 6 shows the regions of the pH-temperature range under which the two generic species of methanogenic bacteria are viable (i.e., the term $EY - k_d$ is positive) at different levels of substrate concentration. Although the viability contour at saturation levels is larger for the hydrogen-consuming bacteria, this does not indicate that they are more robust in general.

D. OTHER FACTORS

The effects of oxygen and toxins were also considered, but the present model is too simple to include their effects in a consistent manner—for instance, the facultative anaerobic bacteria, which would provide a mechanism for the removal of trace oxygen, were not modeled. The E_h level is assumed to be favorable for methanogenesis, as a consequence of the statement that the facultative anaerobes have removed all the oxygen, and so E_h has not been included as a parameter. Accordingly, it is assumed that no oxygen is present in the system and there are no unusual toxins other than those normally present (the effects of which are implicitly incorporated in the growth yields and reaction rates).

VII. pH-REGULATING MECHANISMS

The pH of the liquid within a landfill exerts a major influence on most of the processes occurring therein, though the mechanisms are often obscure and indirect. If we consider a solution containing only the organic acids shown in Figure 1, then there is no substrate level at which the methanogens are viable—since, as the acetate concentration increases, the pH decreases rendering the bacteria inactive. Ammonia is the only major alkali

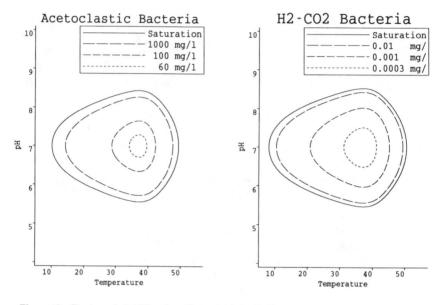

Figure 6   Region of viability of methanogenic bacteria.

produced during the degradation of household waste but it is not present in sufficient quantities to raise the pH to neutral levels.

The results of the simulations together with analyses of leachate samples indicate that the crucial stabilizing influence is the presence of metallic cations which are generated from the biologically inert materials. During the initial anaerobic phase, large amounts of organic acids within the leachate act on the metallic and mineral fractions of the waste, causing a rapid release of cations and, thereby, retarding the decrease of the pH. The main metallic ions found are Na^+, K^+, Ca^{2+}, and Fe^{2+} (the latter being present only in relatively sulfur-free systems) typically in concentrations of 100 to 2000 mg/dm^3 each.

Sodium (and likewise potassium, which can be considered with it because of their similar properties) is extremely soluble, so that once its ions are formed it is unlikely to be precipitated. Sodium alone cannot be responsible for the observed buffering since a concentration high enough to neutralize the peak acid levels would probably create an initial highly alkali environment—and, even if this was not lethal to the bacteria, then once methanogenesis was established at a near-neutral pH, the residual acid concentration would be higher than normally observed.

The solubility of calcium varies drastically over the pH range 4 to 8 (assuming it is in equilibrium with solid calcium carbonate and gaseous carbon dioxide at 50 kPa) and increases as the pH decreases so that large quantities can dissolve to counteract high acid concentrations. However, by itself, it is too insoluble above pH 6 to produce the final neutral leachate observed.

Accordingly, the model suggests that the landfill ecosystem requires both "permanent" sodium buffering (which persists at neutral pH) together with "variable" calcium buffering (which increases enormously as the leachate becomes more acidic) if the methanogens are to prosper. This pH-dependent nature of the cation concentration greatly enhances the homeostatic influence of the bacteria and allows them to maintain the leachate in a near-neutral state while still having adequately high levels of acid present to act as substrate. The model incorporates this dual buffering mechanism by assuming that there are finite reservoirs of sodium carbonate and calcium carbonate present in the waste (possibly being supplied by building materials and soil). The sodium carbonate is assumed to dissolve at an exponentially decreasing rate, and the rate of the calcium solution/precipitation reaction is taken to be such that any imbalance decreases with a specified half-life in the absence of other chemical changes.

VIII. AN EXAMPLE SIMULATION

The final set of equations was integrated numerically using a Fortran programme employing backward differentiation (Gear's Method) and forward differencing.

To demonstrate the type of results obtainable using this model, consider a particular case. Take a landfill composed of waste having initially $150g/dm^3$ carbohydrate, $30g/dm^3$ protein, $20g/dm^3$ fat, and $450g/dm^3$ of nondegradable materials, with each of the putrescible compounds divided in the ratio 1:11:8 between the rapid, medium, and slow decay rate categories. The initial dissolved and gaseous concentrations of CH_4, H_2, O_2, H_2S, NH_3, acetic acid, carboxylic acids, alcohol, and glucose were taken as zero with the atmosphere 20% CO_2 and 80% N_2 (a reasonable starting point under the assumption that aerobic reactions had just finished depleting the oxygen). A population of 10^{-2} mg/dm^3 of each of the methanogenic species was assumed to be resident at the commencement of the anaerobic phase.

The fractional water input was taken to be 10^{-5} $dm^3/dm^3/h$ of clean water, with an equal outflow of leachate. The half-life for calcium equilibration was taken as 6 h. It was assumed that the inert waste contained 500 mg/dm^3 of sodium, and that this passed into solution with a half-life of 100 h, ultimately giving a dissolved concentration of 1650 mg/dm^3 within the leachate.

The leachate's pH during the 14-month simulation is shown in Figure 7. Figure 8 records the levels of acid, acetate, and alcohol/sugar in the leachate, with Figure 9 illustrating the corresponding methanogenic populations. Note that the void fraction occupied by water is 0.3, so that the

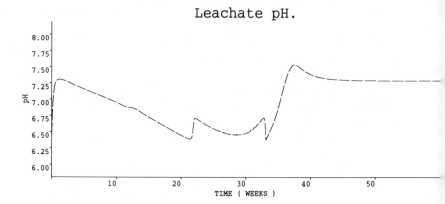

Figure 7   Leachate pH during simulation.

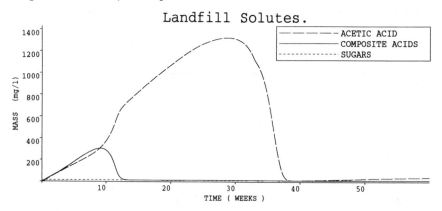

Figure 8   Dissolved concentrations of sugar/alcohol, acetic acid, and carboxylic acids.

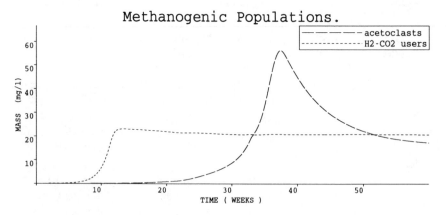

Figure 9   Populations of methanogenic bacteria.

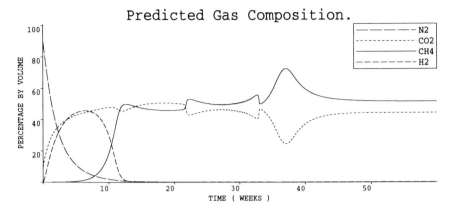

Figure 10   Partial pressures of landfill gas during simulation.

actual leachate concentrations are 3.3 times the mass/dm^3 illustrated in Figures 8 and 9.

Phase plane analysis predicts that a system consisting of only one species of bacterium with activity obeying Monod's Law, together with its substrate which is being produced at a constant rate, has a single non-zero critical point, which is a stable node. This behavior seems to carry over to the more complex system.

Figure 10 shows the daily averages of the partial pressures of the gas being vented from the model landfill.

The gases initially present inside the site are flushed out within the first 10 weeks, as can be seen from the decrease in the partial pressure of nitrogen.

At about 16 weeks the population of H_2/CO_2-consuming bacteria has grown large enough to assimilate all of the hydrogen being produced, and the gas is no longer vented in significant quantities. At the same time, the first methane production is observed.

By week 30 the population of acetoclastic methanogens has increased sufficiently so that it can metabolize all of the acetate being produced and make inroads on the amount stored in the leachate. The large "food reserve," in the form of dissolved acetate, causes the population to overshoot, and a slow decline to a stable steady-state population is observed in the subsequent period.

The rapid rise in pH caused by the removal of acetic acid from solution greatly increases the solubility of carbonate, and there is a short period when the partial pressure of CO_2 is reduced due to additional carbonate being retained in the liquid. Although both CH_4 and CO_2 production are greatly elevated during this short time, the apparent alteration in the ratio of their partial pressures is entirely due to the solubility effect and is not caused by biological mechanisms.

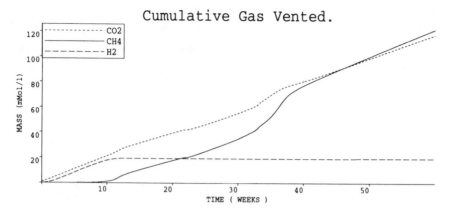

Figure 11   Total vented metabolic gases.

Figure 11 gives the cumulative molar masses of methane, carbon dioxide, and hydrogen which have been vented during the simulation.

IX. DISCUSSION

In all cases where both species of methanogens (i.e., the acetoclasts and the hydrogen consumers) are able to survive, they eventually reach an equilibrium population where substrates are consumed as fast as they are produced, and the degrading waste is converted to carbon dioxide and methane in a roughly 1:1 molar ratio.

In some actual mature sites a gas composition of more than 50% methane is sustained over a period of years. An overall mass balance shows that carbon dioxide is never generated in lesser quantities than methane from carbohydrates, with a small excess of methane being produced during fat degradation (provided the process goes to completion). Examination of the simulated leachate concentrations reveals that a fraction of the carbonate is being stored in the leachate to provide buffering capacity, and another portion is flushed away as water passes through the system. The relative insolubility of methane results in very little of it being involved in these physical processes, so that most of it is out-gassed immediately.

The model predicts a higher concentration of hydrogen during the early phase than is typically observed (the partial pressure in real sites would not normally be expected to exceed 20%) but with spatial variation this may disappear due to averaging as different sections of the site pass through the hydrogen generating phase at different times—however, anecdotal evidence describes some sites as being capable of generation up to 40% hydrogen for short periods. The rate equations (13, 18, and 19) are oversimplified and one would like to replace them by explicit bacterial (or fungal) populations when sufficient data become available.

Diffusion limitations within the liquid mean that the levels of dissolved gases present in the water held within the smaller waste pores (e.g., soaked into paper) may be well above the equilibrium values predicted from Henry's law. This dissolved concentration is highly relevant to bacterial metabolism but until experimental evidence is produced the magnitude of any excess cannot be estimated. The volume of water inflow/outflow prescribed in the above simulations had a negligible effect on the landfill reactions.

Increasing the amount of readily degradable waste above that assumed in the example simulation does not speed up the establishment of methanogenesis since the premethanogenic substrate levels already greatly exceed the Monod half-velocities after the first few days. Indeed, if there is too much readily degradable waste present, then the initial acid production will be so high that the methanogens are killed (or rendered inactive) by low pH before the population has time to increase.

At some landfills taking pulverized waste that has gone through an initial aerobic phase, it has been observed that methanogenesis is established faster, and it could be that the high rate of aerobic degradation (which does not yield acids) removes much of the readily degradable substrate and prevents damagingly high initial acid concentrations in these sites. Similarly, the higher initial temperature may reduce the disparity between the rates of development of the methanogens and the other degradative bacteria (see the comments below on thermal regulation).

Field observations generally do not distinguish whether methanogenesis from acetate cleavage or from carbon dioxide reduction comes first, and it is commonly accepted that both processes start at about the same time. The two modeled bacteria are partially coupled in that an increase in the consumption of one will speed up the composite acid degradation and produce more substrate for both species. An alternative explanation for the apparent simultaneity of both forms of methanogenesis lies in the ability of some species to derive energy both from acetate and hydrogen via Reactions 7 and 8. Such bacteria could develop rapidly on the more favorable substrate, then when this becomes depleted there will be a large population available to consume the other substrate. Before this effect can be modeled some basic microbiological research on the switching mechanisms and critical concentrations will be necessary. Each doubling of the size of the initial seed populations (up to about 0.1 mg/dm^3) results in a linear decrease in their establishment time.

Once the bacteria are firmly established, they maintain the pH at about 7.4, which is fairly typical of a site during the methanogenic stage. During simulations in which the temperature or the amount of sodium present was reduced, this final value decreased toward neutrality. Low metal concentrations are frequently associated with a failure of methanogenesis to develop, and this has been attributed to some of the metals being essential nutrients for the bacteria. However, the model implies that there is another effect in that low metal concentrations mean the leachate is poorly buffered against the carboxylic acids so that the pH may fall too quickly for the methanogenic population to reach a self-regulating size. These pH-raising effects of sodium, potassium, and calcium may be more important than their nutritional aspects, since the levels required for adequate buffering are above those at which the bacterium's nutritional requirements are satisfied. Some experimental work is required to assess the availability of these metals within waste and the rate at which they come into solution—which, if the metals are initially bound within the organic fraction of the waste, will be affected by the amount of degradation which has taken place, and perhaps also by any initial aerobic phase.

The transitions between periods of near steady-state conditions are more sudden than found in actual landfill observations, and this is probably caused by the neglect of spatial variations under the "well-mixed" approximation. In real sites the initial methanogenic populations would be distributed unevenly so that separate regions tend to pass through each growth stage at different times, with diffusion further smoothing the process. The heterogeneous nature of the landfill medium provides a large number of microenvironments, some of which will be more suitable for methanogenesis than an average well-mixed system, and this makes the reactions more robust than our model indicates.

A. THERMAL REGULATION

Another aspect of the Oxford model concerned thermal regulation of sites and showed that the effect of anaerobic degradation on site temperature is very small over the periods (up to a year) in which methanogenesis begins. This information gives confidence that the neglect of thermal feedback in the main biological model does not affect the results obtained using it, nor the conclusions drawn therefrom.

B. MUTATION

To investigate the possibilities of a single species of methanogens being able to metabolize both acetate and hydrogen, a constant mutation rate was postulated at which the acetoclasts and hydrogen-users acquired the alternative methanogenic potential. The effect of this was primarily to handicap the faster growing species, with relatively little benefit to the slower growing type (the two species alternately occupied either role according to the initial conditions). This behavior seems detrimental to the methanogens, and the conclusion drawn is that if the same species is able to use both substrates then its preferred metabolic pathway must be dependent on the actual concentrations of substrate present.

C. AEROBIC PREDIGESTION

As discussed above, increased volumes of highly degradable waste do not promote the onset of methanogenesis. In contrast, refuse pulverization prior to landfilling reduces the lag period for methanogenesis by a combination of substrate removal and elevated temperature (and, possibly, increased availability of metals). Some experimentation is, therefore, required here, as it may be beneficial to adopt a policy of exposing some or all of the landfill waste to oxygen before infilling.

X. CONCLUSIONS

When subjected to a variety of initial and "boundary" conditions the model responded in a sensible fashion and eventually settled down to some steady state, though without methanogenesis under the harsher starting regimes—such as those with too low a temperature or insufficient metal buffering. Given the amount of variance between real landfill sites, the behavior of this model falls well within the range of observed characteristics.

The lack of data which prevent fully incorporating feedback effects into the pH and temperature means that the model has little predictive power regarding primary degradation—it makes no claim to estimate the total gas production potential—however, the simpler nature of methanogenesis means that useful information can be extracted about the interdependencies of the latter processes.

It would be highly beneficial to conduct further experimental research on the metal leaching, temperature, and gas evolution functions. It may be useful to compare sites that are rich/poor in calcium and do/do not contain refuse which has undergone aerobic precomposting. Additional modeling work would be valuable to explore the effects of spatial dependency on isolated subsets of the ecosystem, investigating such things as the rate at which a methanogenic region can spread to adjacent areas.

Other aspects of the Oxford model (not included here) indicated that the composition and volume of vented gas can exhibit large fluctuations even though the internal generation rates are steady, one cause being changes in atmospheric pressure (this is now being borne out by site observations). Although not incorporated into the biological model, these findings explained many of the features of landfill performance previously ascribed to rapidly varying internal fauna, and reinforce the author's view that microbiological explanations of landfill phenomena must take into account the whole local ecosystem if they are to provide robust predictions.

XI. MAIN NOTATION

Below are listed the variables used in the Oxford mathematical model.

XII. NOTATION FOR THIS CHAPTER

k	Decay factor for gas evolution/solution rate, /s
Q_i	Mass rate at which molecules of gas i are evolved, kg/m³/s
t	Time, s
u	Darcy water velocity, m/s
v	Net Darcy gas velocity, m/s
z	Depth coordinate, positive downward, m
C_i	Dissolved concentration of solute i, kg/m³
D_{ij}^b	Binary diffusion coefficient of gases i and j in free space, m²s
D_{ij}	Modified binary diffusion/dispersion coefficient of gases i and j, m²/s
E_i	Combined diffusion and dispersions coefficient of gas i, m²/s
E_i^w	Combined diffusion and dispersion coefficient of solute i, m²/s
H	Henry's constant, kg/m³/Pa
$\underline{\underline{K}}$	Permeability of site to gas flow, m²/s/Pa
$\overline{\overline{K}}_v$	Vertical permeability of site to gas flow, m²/s/Pa
P	Total pressure of gas, Pa
P_i	Partial pressure of ith gas, Pa
Q	Total rate of gas evolution, kg/m³/s
S_i	Rate at which molecules of gas i are created by biochemical activity, kg/m³/s
$T_{1/2}$	Half-life for gas evolution/solution, s
V	An arbitrary volume within the landfill
∂V	Surface surrounding volume V
λ_i	Conversion factors from gas densities to pressures, Pa m³/kg
ϕ	Void fraction occupied by gas
θ	Void fraction occupied by water
ρ_i	Density of ith gas, kg/m³
τ, τ^w	Tortuosities for diffusion in gas and water phases
Ω	Rate of gas evolution/solution, kg/m³/s

REFERENCES

1. Archer, D. B. and Fielding, E. R., *Effluent treatment and disposal*, I, *Chem. E. Symp. Ser.*, 96, 1986, 331.
2. Hoeks, J., Significance of biogas production in waste tips, *Water, Air Soil Pollut.*, 1, 323, 1983.
3. Coleman, P. F., Gupta, A. K., and Oldham, W. K., in *Proc. Int. Symp. New Directions and Research in Waste Management and Residuals Management*, Jasper, S. E., Ed., University of British Columbia, Vancouver, 1985, 376.
4. Huang, J.-C., Huang, Y.-J., and Ray, B. T., in *Proc. Int. Symp. New Directions and Research in Waste Management and Residuals Management*, Jasper, S. E., Ed., University of British Columbia, Vancouver, 1985, 582.
5. Findikakis, A. N. and Leckie, J. O., Numerical simulation of gas flow in sanitary landfills, *J. Environ. Eng. Div.*, 105, 927, 1979.
6. Archer, D. B. and Harris, J. E., *Anaerobic Bacteria in Habitats other than Man*, Barnes, E. M. and Mead, G. C., Eds., Blackwell Scientific, Oxford, 1986, 185.
7. Pirt, S. J., *Principles of Microbe and Cell Cultivation*, Blackwell Scientific, Oxford, 1975.
8. Lawrence, A. W. and McCarty, P. L., Kinetics of methane fermentation in anaerobic treatment, *J. Water Pollut. Control Fed.*, 41, R1, 1969.
9. Farquhar, G. J. and Rovers, F. A., Gas production during refuse decomposition, *Water, Air Soil Pollut.*, 2, 483, 1973.
10. Watson-Craik, I. A. and Goldie, S., unpublished data, 1990.
11. Zinder, S. H., Microbiology of anaerobic conversion of organic wastes to methane: recent developments, *ASM News*, 50, 294, 1984.

CHAPTER **4**

Co-Disposal of Industrial Wastewaters and Sludges

Irene A. Watson-Craik and Kevin J. Sinclair

CONTENTS

0-87371-968-9/95/$0.00+$.50
© 1995 by CRC Press, Inc.

I. INTRODUCTION

Co-disposal is generally defined, in the U.K., as the disposal of industrial wastewaters and sludges with domestic, commercial, and industrial refuse in a class I or class II landfill site, i.e., those sites in which leachate is collected for treatment or which rely on leachate attenuation processes within the substratum.

While industrial wastes need not *per se* be hazardous, many are generally considered toxic or hazardous. The term "hazardous" has, however, proved difficult to define legally. In the U.K., hazardous waste is not defined in the legislation but many industrial wastes, which are considered under Section 17 of the U.K. Control of Pollution Act (1974) to be "dangerous" or "difficult to dispose of", are referred to as "special wastes." The U.K. Control of Pollution Act (Special Wastes) Regulations (1980), which were a transport manifest only and were aimed to make provision for tighter control over the carriage of wastes which could seriously threaten life, further defined special wastes. By this, inclusion was made for any substance listed in Schedule 1 of the Regulations and which was dangerous to life by way of corrosivity, toxicity, or carcinogenicity; medical products available only on prescription; and listed products of low (<21°C) flash point. Schedule 1, however, has some notable omissions such as, for example, arsenic metal (which can emit toxic arsine on acidification), azides, and metal hydrides. Moreover, it ignores damage to plant and animal life, water pollution, and site sterilization.[1]

In the U.S., where the approach is more definitive, the Environmental Protection Agency (EPA)[2] defines a waste as hazardous if

  1. It exhibits ignitability, corrosivity, reactivity, or extractive procedure toxicity, as determined by standard tests.
  2. It contains any of the toxic constituents named on published lists as having toxic, carcinogenic, mutagenic, or teratogenic effects on human or other life forms.
  3. It is listed on prescribed lists.

These prescribed lists name several hundred specific chemicals and "waste streams", including known or suspected carcinogens, pesticides, heavy metals, and acids. An equivalent list published in the U.K. by the DOE[3] was constituted by 174 groups of wastes in 18 categories and included organic acids, toxic metal compounds, organic compounds, and animal and feed wastes. As in the U.S., this list is subject to continual amendment and expansion.

European Council (EC) policy development has similarly been hampered by a lack of a mutually agreed definition of what constitutes a hazardous waste.[4] The amendments of the Council Directive of December 12, 1991, on hazardous waste (Annex 2), which came into force in December 1993, doubled the number of materials classified as toxic. Previously, materials were deemed hazardous if they contained elements of 1 or more of 27 substances. The amended directive not only doubles the number of contaminants but includes products generated by listed activities, presuming in either case that the material also exhibits 1 of 15 physical properties, such as a high degree of flammability or toxicity. Brand[5] noted that, despite its comprehensiveness, this definition has inherent problems due to the fact that there is a gray area between the concept of ''waste'' and ''recyclable material'' that cannot be legally defined. The exact boundary between the two is continuously redefined by changing market forces. In other words, a rigid, process-based definition of hazardous waste, as favored by the U.S. and the EC, would seem to guarantee its safe disposal but could reduce its chances for recycling. Laurence and Wynne[6] doubted whether, in fact, a uniform definition which is appropriate to all circumstances could ever be developed. In their view, ''the wide, almost infinite, variety of circumstances involved in waste handling and trading militates against a single water-tight system of definition.''

One caveat to be sounded on adherence to definitions and lists is that they confine the problems within specific limits. It has been pointed out,[7] however, that all materials can present a danger under certain conditions. Thus, for example, even sodium chloride, which is essential to life, is corrosive and can disturb the electrolyte balance in man. It is perhaps pertinent to point out that the converse is not true since a hazardous chemical must be considered a potential danger under all circumstances. Although it has been suggested[8] that ''the body has a natural ability to resist unwanted and dangerous materials if at low enough concentrations and, therefore, they can be regarded as safe,'' such a statement utterly disregards the risks of bioaccumulation and possible chronic effects of long-term exposure to low concentrations. Prescribed lists also fail to recognize hazards arising from catalytic or antagonistic reactions of two or more substances which are in themselves harmless. Moreover, the real threat posed by a material may bear no relation to the threat as perceived by the general public. Thus, disposal is a balance not only of technology and economics but also of politics.

II. GENERATION AND DISPOSAL OF HAZARDOUS WASTES

Total waste production in the EC has been estimated at approximately 2500 million tonnes per annum, 1% of which is currently defined as hazardous.[5] The Organization for Economic Cooperation and Development estimated in 1985[9] that 10% of all hazardous waste generated in Europe crosses at least one national border before its final disposal, and the implications of the 1992 "single market," in terms of waste treatment and disposal, have still to be critically assessed. These cross-frontier shipments partly result from local capacity shortfalls but also from regional differences in the regulatory standards for disposal. The legal disposal of hazardous waste in a foreign landfill may be cheaper than incineration in the country of origin, transport costs notwithstanding. Thus, imports of toxic/hazardous waste into the U.K. have increased from 35,000 tonnes in 1989/90[10] to 170,000 tonnes in 1992.[5]

For both the U.S. and the EC the annual increase in waste generation is between 2 and 4%.[11] The close correlation observed[12] between gross domestic product (GDP) and general waste arisings motivated the suggestion that forecasts could be made based on assumptions of the likely rate of increase of GDP. The impact of this growth is not only environmental but also economic since at the same time as companies are having to face increasing costs for disposing of their wastes revenues of waste disposal companies will increase.

These high costs of waste disposal and the potential long-term legal liabilities are likely to provide the necessary economic incentive to increase the disparities between the volumes of hazardous wastes generated and those requiring disposal.[11,13,14] At present, in the U.K., typical of the industrialized nations, the largest proportion (65%) of hazardous waste generated is recycled or reclaimed by industry.[13,15] The countries of the EC were estimated, in 1988, to produce approximately 20 million tonnes per annum (Table 1) of which approximately 35 to 47% required disposal.[14]

While manufacturers are increasing their efforts to minimize waste production, to reclaim or recycle hazardous materials, or to find a customer who can use the materials, large volumes of waste require, and will continue to require, safe disposal. Fortunately, there is a wide range of technologies applicable to the disposal of hazardous wastes, including thermal (pyrolysis and incineration), chemical (e.g., oxidation, precipitation, ion exchange, microwave plasma technology), physical (including carbon sorption, dialysis, filtration, flocculation), and biological (e.g., activated sludges, trickling filters).[10,16,17]

Despite this multiplicity, most waste is disposed of to landfill in the U.K. (Tables 1 and 2). The use of incineration has been limited on a cost basis[17] and, additionally, it is not suitable for many types of wastes and

Table 1 Hazardous Waste Generation in the European Community

Member state	Hazardous waste generated (× 1000 tonnes)	Disposal infrastructure (×1000 tonnes)		
		Incineration	Physical-chemical treatment	Landfill
Belgium	1,000	1	1	4
Denmark	67	1	1	1
France	4,300	25	10	12
The Netherlands	1,000	7	6	?
West Germany	4,500	17	23	22
United Kingdom	3,750	4	13	1,200
Greece	—			
Ireland	76			
Italy	2,000–5,000			
Luxembourg	—			
Portugal	—			
European total	19,943			

Adapted from Mettelet, C., Production, traitment, recyclage et transferts transfrontaliers de déchets dangereux dans la Communauté Européenne. Paper presented at the STOA Workshop "Hazardous Management Beyond 1992", Brussels, April 25–26, 1989.

Table 2 Disposal of Special Wastes in the U.K.

Treatment method	Volume (tonnes)	% of total
Direct landfill	1,330,000	70
Physical-chemical	285,000	15
Sea disposal	190,000	10
Incineration	95,000	5
Total	1,900,000	100

Adapted from Department of the Environment, *Digest of Environmental Protection and Water Statistics,* Her Majesty's Stationery Office, London, 1990, chap. 5.

may generate toxic gaseous products in conjunction with unburned particles. At present, since return of investment invariably favors product research and development to the exclusion of pollution control,[18] more costly means of disposal will only be employed when cheaper means have proved ineffective.

Co-disposal has been endorsed in the U.K. by the Department of the Environment (DOE),[19] by the House of Lords Select Committee on Science and Technology,[13] which considered that "for many liquid industrial wastes co-disposal with domestic refuse, if well executed, is . . . a valid method, which waste disposal authorities should be encouraged to adopt more readily," and by the Hazardous Waste Inspectorate,[15] who could see ". . . no objection to more co-disposal, if well managed, as the BPEO (Best Practicable Environmental Option)." As a result, and with the retention of co-disposal in the Draft EC Directive on landfilling, co-disposal is likely to remain a widely utilized disposal method for hazardous wastes for the foreseeable future.

Co-disposal is, however, at present discouraged or outlawed in many countries such as Germany, Canada, Australia, and the U.S. due, principally, to previous co-disposal practices and the resultant adverse environmental impacts and public opinion backlash. Incidents such as those at Love Canal (Niagara, NY) have achieved international notoriety. More than 19,000 tonnes of unspecified liquid and solid chemical wastes were deposited in this site between 1942 and 1953. In 1978, a year after a survey had detected the migration of chemicals into nearby basements, the New York State declared a Health Emergency and evacuated 235 families. By 1980 relocation had been offered to all local households. At least 248 chemical compounds, including 35 neurotoxins, 34 carcinogens, 20 hepatoxins, 15 renal toxins, and 30 embryo toxins, were identified in Love Canal leachate. Health surveys detected adverse effects on the health of children, with increased incidences of seizures, learning difficulties, hyperactivity, skin rashes, and incontinence, and also on the birth weights of babies born locally.[20,21]

As a result of such incidents, even though they arose as a result of practices now unacceptable, legislation in the U.S. demands that all new sites be constructed with a liner and facilities for leachate collection, and that the disposal of hazardous wastes occurs only in dedicated sites. Similar legislation is implemented in Germany.

It is particularly unfortunate that public confidence has been sacrificed as properly managed co-disposal could offer several advantages. As detailed above, disposal to landfill is generally cheaper than other disposal methods. Moreover, the increased operation of landfill sites for co-disposal could also minimize transport costs. In 1989 and 1990, for example, it was estimated that 10,668, 5757, 5734, 4237, and 1853 tonnes were imported into the U.K. from The Netherlands, Belgium, Switzerland, U.S., and Italy, respectively.[10] Risks of environmental damage from tanker leakage or accidents would also be diminished.

The operation of a co-disposal site as a multimillion m^3 anaerobic bioreactor could confer advantages similar to those offered by smaller-scale downflow stationary fixed film reactors (DSFFRs). These reactors have proved highly flexible, with successful operation at temperatures between 10 and 55°C and high tolerance of severe and repeated hydraulic and organic overloadings. DSFFRs can tolerate downtimes of weeks or even months, without a great loss in activity, particularly at temperatures of <25°C,[22] although the duration of adaptation of microbial communities to xenobiotic compounds may prove the limiting factor in tolerable downtimes. Spain and van Veld,[23] for example, examined the adaptation of sedimental microbial communities to the degradation of a range of xenobiotics and reported that adaptation to p-nitrophenol was not detected 7 weeks after the initial exposure.

Rehabilitation of the site surface and co-disposal of liquids via stand pipes to the bioreactor beneath offers aesthetic advantages over, for example, an incineration plant. The infiltration of liquids may, moreover, reduce the period required for site stabilization,[24] although Rees[25] considered that rapid infiltration of water may result in excessive cooling and the subsequent inhibition of methanogenesis.

The economic extraction of methane from landfill sites may also be enhanced by liquid co-disposal practices. The correlation between refuse moisture content and rates of gas production was reviewed by Rees,[26] while Kasali et al.[27] observed maximal total evolution of methane at a moisture content of 75% (w/w), which approximated to the field capacity. In the U.K., freshly emplaced refuse has been reported with moisture contents from 25 to 35% (w/w)[28-30] or 19 to 32% (v/v),[31] which are well below the optimal values for methane production.

Preeminent, however, is the operation of a co-disposal site for the safe and effective disposal of specified industrial wastes. Three targets should be achieved:

1. There should be no impairment of leachate quality.
2. There should be no added risk to ground or river water from the increased volumes of liquid added.
3. Public opinion should be satisfied. Khan[32] pointed out that past practices had resulted in public pressure against the landfilling of waste and that a strong public relations exercise was necessary to gain public confidence in both the waste disposal industry and the regulatory authorities.

To accomplish these targets necessitates a thorough understanding of attenuation mechanisms operative for specific wastes, of pertinent hydraulic and organic loading rates, and of the interdependent effects of landfill metabolism/added xenobiotic. As co-disposal is, fundamentally, a superimposition onto landfill catabolic processes, effective co-disposal must be assessed in terms of these. To date, however, development of effective co-disposal strategies has been constrained by several factors.

The first factor is the wide range of industrial wastes disposed of to landfill in the U.K., which includes arsenical wastes, where 80% (9 × 106 t/y) are deposited in landfills;[33] asbestos, mercury, and acid wastes;[19] and wood preservatives.[34] Not only do most industrial wastes comprise several categories of hazardous wastes (Table 3), but also landfill sites are generally licensed to accept hazardous wastes of more than one type. An example of this was given by Cheyney,[35] who detailed the categories and quantities of >90 notifiable wastes deposited over 18 months in one landfill site. These included phosphoric acid (579 t), organic acids (26 t), zinc (500 t), cyanides (25 t), asbestos (182 t), phenols (1 t), herbicides (27 t), tannery wastes (2908 t), and food processing wastes (171 t). In addition, the contribution of domestic refuse to the production of hazardous wastes is often overlooked. It

Table 3   Representative Hazardous Substances within Industrial Waste Streams

Industry	Arsenic	Cadmium	Chlorinated hydrocarbons[a]	Chromium	Copper	Cyanides	Lead	Mercury	Selenium	Zinc	Misc. organics[b]
Battery		+									
Electrical/electronic			+	+	+	+	+ +	+ +	+	+	+
Explosives	+			+ +	+ +	+	+ +	+ +			+
Leather				+ +							+
Paint/dyes		+	+	+ +	+	+ +	+ +	+ +	+		+
Pesticides	+					+ +	+ +	+ +		+	+ +
Coal conversion					+	+				+	+ +
Pharmaceuticals	+				+			+			+ +
Textile				+	+						+ +
Pulp/paper								+			+

[a] Includes polychlorinated biphenyls.
[b] For example, aniline, phenol, chlorophenols, dinitrobenzene.

Adapted from Fuller, W. H., *The Scientific Management of Hazardous Waste*, Cambridge University Press, Cambridge, 1983, chap. 8; Fedorak, P. M. and Hrudey, S. E., *Water Sci. Technol.*, 17, 143, 1985.

is estimated that in the U.S. each household contributes each year to a municipal landfill >4.55 l of hazardous waste,[36] which is neither specified nor controlled.

This multiplicity of substrates, many of which are likely to be present at low concentrations, poses specific problems. For example, it was originally noted by Boethling and Alexander[37] that with selected molecules anomalous patterns of biodegradation may be observed at low concentrations. For example, although two thirds of the C2 atom of the acetate moiety of 2,4-dichlorophenoxyacetic acid were mineralized at 22 µg/ml and 220 ng/ml, <10% was converted to CO_2 at 2.2 ng/ml or 22 pg/ml. It was suggested that energy is obtained too slowly from oxidation of the low substrate concentrations to meet the energy demands of the initially small population active on the compound. This hypothesis was supported by the observation of Pahm and Alexander[38] that the addition of 10 ng of glucose per milliliter resulted in mineralization of *p*-nitrophenol at concentrations too low to be mineralized when the nitro compound was the sole source of added carbon, and that bacteria may thus be able, in natural environments, to mineralize compounds at concentrations below those able to support growth in culture media because of the availability of other carbon sources. Degradation patterns such as these may be particularly important in the landfill ecosystem, which is characterized by a wide range of substrates, many at low concentrations.

Rittmann[39] considered the effects of substrate interactions on biodegradation and noted that primary substrates, essential for growth and maintenance, also influence the degradation of specific hazardous molecules by being inducers, inhibitors, and/or direct or indirect co-substrates. The target xenobiotics also have the interactive roles of self-inhibitor, inhibitor of primary substrate utilization, inducer, and a part of an aggregate primary substrate. For example, Saez[40] noted that degradation of 4-chlorophenol by *Pseudomonas putida* was induced by the presence of phenol. Few data are available, however, on possible substrate interactions under the environmental conditions (E_h, temperature, pH, moisture, etc.) and nutrient concentrations characteristic of landfill sites.

The second factor is the absence of information on the possible hazards of mixing wastes on sites. Cook[7] predicted that chemical reactions could take unexpected paths if chemicals were in proximity for long periods in the presence of oxidizing agents and if catalytic materials were present. It is important, therefore, that each sample is analyzed prior to disposal and that wastes known to be incompatible are segregated. Long-term problems may also arise if accurate records of previous disposals are not maintained. This risk would be minimized if biodegradation reduced the concentration of the waste or mineralized it completely.

The third factor is a paucity of research on microbially mediated degradative processes in refuse. Where the waste is biodegradable, or potentially so, the long-term goal must be detoxification of the waste by complete or partial biodegradation. Senior,[41] however, pointed out that, perhaps due to the complexity and heterogeneity of the landfill ecosystem, in addition to increasing amounts of wastes requiring disposal and to the decreasing number of suitable sites in the right places, research has focused on the civil engineering aspects of landfill technology and fundamental microbiological and biochemical studies have largely been neglected. Information on loading rates, for example, is almost completely lacking and the WMP Landfill Practices Sub-Group 2[42] noted that, even where values did exist, these seemed quite arbitrary and derived on the grounds of "what has been done in the past and caused no problem is probably acceptable for future operations." There are several flaws in this philosophy, not the least of which is that problems arising from past operations may not yet have become apparent.

The fourth factor is optimization and exploitation of the refuse fermentation. These operations necessitate not only a fundamental understanding of the indigenous microbiology and biochemistry but also assessment of the fermentation performance. There is at present no consensus on the selection of indicator parameters although there are several whose selection as indices of landfill stabilization has been frequently reported. These, generally, necessitate the analysis of leachate and their usefulness is thus diminished in recently opened sites or in those sites where ingress of liquid is discouraged.

Landfill age and, hence, the degree of waste stabilization has a significant effect on the composition of leachate.

Leachate pH was reported to increase the time, within the range 5.2 to 8.3.[43] This was corroborated by Robinson and Maris[44] who reported leachate pH values of 6.21 and 7.45 from wastes emplaced for <2 and >20 y, respectively. These differences in leachate pH, in general, reflect the variations in total volatile fatty acid (VFA) concentrations, and Harmsen,[45] for example, noted that a leachate (pH 5.7), from a site at the acidification stage, contained concentrations of acetate, propionate, butyrate, and hexanoate of 11,000, 3760, 9890, and 5770 mg/l, respectively. Conversely, no VFAs were detected in leachate (pH 7.0) from a site at the methanogenic stage.

In sites where refuse has been emplaced for longer periods, not only are the biochemical oxygen demand (BOD), chemical oxygen demand (COD), and total organic carbon (TOC) contents of the leachate lower,[46] but also the ratio of COD:TOC tends to decrease. This ratio was reported to vary from 3.3 for a relatively young landfill to 1.16 for an older site,[47] which reflected the higher proportion of semi-recalcitrant/recalcitrant compounds such as humic and fulvic acids in the leachate from "aged" wastes.

Nitrogen and phosphorus are chemical parameters often used for the determination of nutrient limitation.[47] Although it has been reported that leachate concentrations of sulfate tended to decrease with refuse age,[48] sample variations of other elements such as magnesium, lead, copper, potassium, and sodium were so great that no correlation between refuse age and element concentration could be derived.

Temperature analyses may detect shifts from aerobic to anaerobic metabolism, or vice versa,[49] while methane release may be monitored by gas-phase analyses, or by analyses of solubilized methane.

Some parameters, such as refuse density, field capacity, or buffering capacity, are indices of age rather than fermentation performance. Others, such as leachate odor or color, are more subjective. Moreover, even potentially useful indicator parameters such as leachate pH and VFA concentration or gas-phase analyses are strongly influenced by first-tier variables such as season, climate, rainfall, depth of fill, compaction, and use and composition of cover. In addition, it has been reported that some methods of leachate sampling and storage significantly influence the results of subsequent analyses.[50]

As co-disposal of wastewaters with refuse may be regarded as a superimposition on to landfill catabolism, its efficacy must be judged, at least partly, in terms of its effect on refuse fermentation. Any laboratory assessment of the refuse fermentation process, therefore, necessitates the application of a range of carefully selected indicator parameters, and defined operational conditions and the judicious use of control systems.

III. CO-DISPOSAL

A. EMPLACEMENT

Of the three methods available for the application of liquid industrial wastes—trenching, lagooning, and spraying—trenching is normally the preferred method. The WMP Landfill Practices Sub-Group 2[42] recommended construction of a trench one excavator bucket wide and several meters deep, with at least 3 m of waste present below the trench bottom. When the trench is nearly full, it is back-filled and a new one excavated. Slurries or sludges such as tarry wastes may also be deposited by trenching methods. Although officially discouraged, and considered " . . . as a general principle, an unacceptable practice,"[51] sludges or solid wastes continue to be deposited in drums.

B. ATTENUATION/IMMOBILIZATION OF WASTES

The DOE[19] admitted that in many landfill sites, where co-disposal had been practiced, the three operational targets previously detailed had not been achieved, with resultant impairment of refuse catabolism and leachate quality as indicated by, for example, the persistence of high concentrations

Table 4   Analyses of Leachates Abstracted from Co-Disposal and Domestic
          Refuse Landfills

| Determinant[a] | Co-disposal Sites | | | Domestic refuse |
	Pitsea	Stewartby	East Tilbury	
pH	7.5	7.7	7.1	7.9
Total organic carbon	623	1076	406	260
Chemical oxygen demand	1673	2676	—	804
Biochemical oxygen demand	203	269	66	18
Ammonia nitrogen	551	1390	399	440
Oxidized nitrogen	0.9	0.7	1.9	8
Alkalinity (as $CaCO_3$)	3895	8418	2622	2900
Sulfate	64	182	212	460
Chloride	2810	3187	1737	3000
Sodium	1969	2561	—	2220
Potassium	452	922	315	515
Zinc	0.33	0.65	0.28	0.18
Copper	<0.06	0.08	0.06	0.09
Nickel	0.14	0.27	0.13	0.12
Chromium	0.17	0.20	0.09	0.08
Lead	<0.11	0.11	0.15	<0.10
Cadmium	<0.03	<0.02	0.01	<0.02
Iron	—	7.9	36	3.8
Manganese	—	0.15	2.1	0.41

[a] All analyses, except pH, in mg/l.

Adapted from Harmsen, J., *Water Res.,* 17, 699, 1983; Knox, K. and Gronow, J.,
Waste Manage. Res., 8, 255, 1990.

(8 to 375 mg/l) of phenol in the leachates and boreholes of several sites.
At one site, the deposition of substantial amounts of acid wastes, particularly sulfuric acid, resulted in pH values of 1.3 to 1.8 and 2.0 to 2.2 in the
aqueous phase of the lagoon and in the saturated zone at the base of the
site, respectively. No microbiological activity was detected in the saturated
zone.

Despite these potential problems, a balanced refuse fermentation presents physical, chemical, and biological mechanisms which can attenuate
many components, either organic or inorganic, of industrial wastes. Knox[52]
argued that, by using this attenuative capacity, the three operational criteria
could be satisfied, and leachate analyses from several co-disposal sites,
including Pitsea (Essex), where co-disposal had been practiced for 20 years,
showed no significant differences in leachate quality from sites accepting
domestic refuse only (Table 4). Leachate quality must, however, be carefully assessed. The use of single leachate analyses to indicate leachate
quality was questioned by Westlake et al.,[53] who noted that constituent
concentrations varied by up to two orders of magnitude within the one site,
and suggested that any differences in leachate composition between types
of site may be hidden by the variation within sites, unless a large number
of samples is taken from each site.

Table 5   Sludge Disposal Options in the European Community

| Country | Land application | Disposal method (% of total) | | | Total (×1000 t dry matter/ year) |
		Landfill	Incineration	Sea dumping	
Germany	25	65	10	—	2750
U.K.	51	16	5	28	1500
France	27	53	20	—	900
Italy	34	55	11	—	800
Spain	61	10	—	29	300
The Netherlands	53	29	10	8	280
Portugal	80	12	—	8	200
Greece	10	90	—	—	200
Denmark	43	29	28	—	150
Belgium	57	43	—	—	35
Ireland	23	34	—	43	23
Luxembourg	80	20	—	—	15

Adapted from Etheridge, S. P., *Environ. Today,* 4, 8, 1993.

C. ORGANIC WASTES

1. Sewage Sludge

Co-disposal has been defined as the disposal of industrial wastewaters and sludges in landfill sites with domestic and other nonindustrial wastes. Sewage sludges are not generally considered industrial wastes, yet they share many characteristics, being produced in large volumes from a (relatively) few distinct point sources. Moreover, it has been reported that the total pollution load resulting from industrial wastewaters is usually at least as great as that from domestic sewage since trade wastes, although comprising <13% of the total volume, are generally much stronger, with higher BOD and COD values.[54] With its high organic content and nitrogen, pathogen, and heavy metal content, sludge disposal presents many of the problems of industrial sludge disposal.

At present, sludge disposal in the U.K. (Table 5) is via three major routes, land application, ocean dumping, and landfilling (co-disposal). Alternative methods of disposal such as incineration,[55,56] composting,[57,58] land reclamation, and forestry[59] are also exploited on a smaller scale. With the commitment by the U.K. to cease ocean dumping by the year 2000,[60] and a rapidly narrowing field of acceptable disposal options,[61] disposal in an economic and environmentally acceptable manner is increasingly problematic. It is generally considered, however, that landfill is likely to remain a major sink for sludge wastes,[61,62] although Kelly[63] suggested that the EC landfill directive, by harmonizing waste disposal practices throughout the EC, will increase the costs of landfilling, and more than half of the 300 presently licensed co-disposal sites in England and Wales are likely to close because they will not meet the requirements of the directive.

Oake[64] suggested that future considerations on the suitability of sludge disposal routes are likely to be influenced by the benefits which may accrue from their use. Specifically, factors such as energy balance and the production of beneficial materials may be important determinants when selecting a particular disposal route. However, informed decisions on sludge co-disposal practices are likely to be constrained by the scarcity of detailed information on the microbial and biochemical transformations which result from the co-disposal of sludge with domestic refuse.

a. Legislation

In the U.K., sludge disposal to landfill is not subject to specific national regulations,[65] although sites that accept sludge wastes must be licensed under the Control of Pollution Act (1974). The main purpose of licensing is ''to ensure that landfilling operations entail no unacceptable risk to the environment and to public health, safety and amenity.''[66]

b. Operation

Whether sludge is added to a preexcavated trench or mixed with incoming solid wastes, minimization of the amount of water entering the site is a primary concern. Sludges generally are dewatered to a dry solids content of at least 25% before landfilling, in order that leachate production is minimized and a reasonable degree of physical stability is maintained. It was calculated[58] that 1 tonne of raw sludge (4 to 8% dry solids w/w) would introduce approximately 990 to 995 l of water compared to only 700 to 800 l if dewatered sludge cake (20 to 30% w/w dry solids) was added. Frost et al.[58] considered that the best practicable environmental option should comprise the following:

1. Disposal limited to containment landfills.
2. Co-disposal restricted to dewatered, digested, and lime-treated primary filter cake sludges.
3. Sludge should either be buried in trenches or spread at the base of the working face and, in both cases, rapidly covered.
4. The ratio of solid waste:sludge should be determined by site water balance calculations.

c. Composition of Sewage Sludges

According to Bruce and Davis,[61] technology is capable of producing at least nine different types of sludge end product, the composition of which will vary according to its method of production and origin (Table 6). Although the effects of adding different sludge types to refuse is likely to yield variable results, this aspect of co-disposal has received little attention,

Table 6   Composition of Selected Sludge Types

Sludge component	Anaerobic	Aerobic	Others
Organic C (%)	18–39	27–37	6.5–48
Total N (%)	0.5–17.6	0.5–7.6	<0.1–10
NH_4-N (ppm)	120–67,600	30–11,300	5–12,500
NO_3-N (ppm)	2–4,900	7–830	—
Total P (%)	0.5–14.3	1.1–5.5	<0.1–3.3
Total S (%)	0.8–1.5	0.6–1.1	—
Pb (mg/kg)	58–19,930	13–15,000	72–12,400
Zn (mg/kg)	108–27,800	108–14,900	101–15,100
Cu (mg/kg)	85–10,100	85–2,900	84–10,400
Ni (mg/kg)	2–3,520	2–1,700	15–2,800
Cd (mg/kg)	3–3,410	5–2,700	4–520
Cr (mg/kg)	24–28,850	10–13,600	22–99,000

Adapted from Bruce, A. M. and Davis, R. D., *Water Sci. Technol.*, 21, 1113, 1989.

with the exception of a study by Craft and Blakey[123] which assessed the effects of different sludge types and mixing strategies. The authors observed that co-disposal increased methane release rates, compared to the domestic waste controls, and incorporation of digested sludge reduced the mass of selected constituents leached from the systems (Table 7 and Figure 1).

d. Site Impact Potential

To date, most studies on the impacts of sludge co-disposal[66-68] have described gas production and leachate quality. However, the microbiological and biochemical bases for observations in these and other studies await clarification.

Harries,[68] for example, studied changes in gas production and leachate chemistry (specifically VFA production) in lysimeters operated with leachate recycle and packed with (a) pulverized refuse; (b) pulverized + composted refuse; (c) pulverized refuse + anaerobically digested sewage sludge; and (d) partially precomposted pulverized refuse mixed with builders' rubble and anaerobically digested sewage sludge. Incubation of the pulverized refuse alone rarely resulted in active methanogenesis, while only limited production, never exceeding 15% of the total gas phase, was recorded in lysimeter (c). Promotion of methanogenesis in lysimeters (b) and (d) was demonstrated by methane concentrations, in the gas phase, of approximately 55%. Coutts et al.[67] measured VFA pool sizes at depths of 5, 10, and 15 m in samples obtained from boreholes in six test cells, and recorded both acetate and butyrate concentrations >20 mM in all cells. However, in a cell which had received sewage sludge, acetate concentrations, at 15 m, exceeded 60 mM.

Most of the concern on the potential environmental impacts of sludges focuses on the following pollutants which may not be contained within the confines of the landfill site.

Table 7   Mass of Material Leached from Co-Disposed Refuse/Sludge Mixtures Expressed as a Percentage of that Leached from Pulverized Refuse Control

Substrate	Refuse:sludge (wt:wt)	Total					
		COD	NH$_4$-N	Phosphorus	Fe	Zn	Ni
Refuse + raw, dewatered sludge	4.1 : 1	−48	+10	+10	−57	−54	−24
Refuse + raw, dewatered sludge (homogeneous mix)	4.1 : 1	−64	+23	+58	−62	−95	−46
Refuse + raw, dewatered sludge (high infiltration)	4.1 : 1	−15	+66	+203	−2	−70	−4
Refuse + primary/mixed dewatered sludge	4.8 : 1	−46	+40	+4	−48	−71	−12
Refuse + liquid, digested sludge	9.7 : 1	−7	−8	−2	+5	−68	−4

Adapted from Craft, D. G. and Blakey, N. C., Codisposal of sewage sludge and domestic waste in landfills, Paper presented at ISWA '88 (5th International Solid Wastes Conference), Copenhagen, September 11–16, 1988.

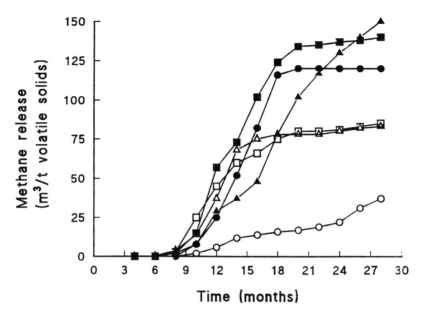

Figure 1   Changes in methane release in reactors packed with specific ratios of refuse:sludge (○) domestic waste only (control):domestic waste + (●) raw, dewatered sludge 4.1:1(w/w); (▣) raw, dewatered sludge (homogenous) 4.1:1 (w/w); (□) raw, dewatered sludge (high infiltration) 4.1:1 (w/w); (△) primary/mixed dewatered sludge 4.8:1 (w/w); and (▲) liquid digested sludge 9.7:1 (w/w). (Adapted from Craft, D. G. and Blakey, N. C., Codisposal of Sewage Sludge and Domestic Waste in Landfills, Paper presented at ISWA '88 (5th International Solid Wastes Conference), Copenhagen, September 11–16, 1988.)

e. Nitrogenous Compounds

Several authors have indicated that landfills may be either nitrogen[69] or phosphorus[70,71] limited. According to Senior[71] the C:N ratio of refuse often exceeds 50:1, compared to the ratio (25 to 30:1) recommended by the Department of Energy[74] as necessary for non-nitrogen limited growth. However, the ratio quoted for refuse may be artificially high, due to the presence of materials, such as lignin, which are nonbiodegradable under the anaerobic conditions of a landfill site.

There may seem benefits, therefore, in the co-disposal of a nitrogen-rich material, such as sludge, since emplacement in landfill sites adds nitrogenous compounds in several forms, the most common of which are NH_4^+/NH_3, organic nitrogen, and NO_3.[72] However, it is generally accepted that sludge co-disposal will not only result in increased volumes of leachate[64] but that these are characterized by elevated ammonia and phosphorus concentrations.[57]

Few studies have addressed the nitrogen fluxes in landfill sites. Most speculation on nitrogen turnover in landfill has focused on the decomposition of proteinaceous material although protein concentrations are generally low (4%) compared to carbohydrates (40 to 50%) such as cellulose.[73] Rees[26] noted that concentrations of the branched-chain fatty acids, *iso*-butyrate and *iso*-valerate, formed as a result of deamination, were low compared to those of the straight-chain fatty acids. Temporal dissociation of proteolytic activity from other degradative processes has been suggested[74] as a result of which easily soluble NH_4^+/NH_3 could be lost in leachate during the early stages of the refuse fermentation, before the microbial population is fully developed. The interstitial fluid within the refuse could, as the fermentation develops, become nitrogen limited.

However, Sinclair[75] noted that significant quantities (>50%) of the organic nitrogen originally present in refuse/sludge mixtures in laboratory-scale lysimeters remained bound within the mixture well into the methanogenic phase of the fermentation, and concluded that a considerable pollution potential appeared to remain dormant within the system.

The fate of nitrate has received little attention. Barlaz et al.[73] assumed that nitrate in fresh refuse samples was utilized solely for carbohydrate oxidation, and Sinclair[75] considered that the possibility of dissimilatory reduction to ammonia should not be overlooked.

Results of studies such as these highlight our present ignorance of the processes controlling nitrogen fluxes in landfill sites, and the relative significance of processes such as assimilatory and dissimilatory nitrate reduction, denitrification, and deamination. However, Sinclair[75] concluded that, with the increased adoption of class 1 (lined) sites and the simplification of leachate collection and treatment, the benefits likely to accrue from increased gas production should exceed the problems, such as elevated leachate nitrogen concentrations, generated as a result of sludge co-disposal.

f. Microbiological Content

Human feces characteristically harbor large numbers of a range of microorganisms (Table 8) which are also recorded in sludges. As a result, it has been speculated that sludge addition to refuse may influence the refuse fermentation by adding significant quantities of key microbial groups/species,[76,77] although Barlaz et al.[73] considered that all the bacterial physiological groups required for refuse methanogenesis were present in fresh refuse. In particular, additions of anaerobically digested sludge may be effective, due to the introduction of an active population of anaerobic bacteria.

Sludges are also characterized by a variety of pathogenic organisms such as viruses, bacteria, and parasites, which will vary in numbers and

Table 8 Numbers of Viable Bacteria in Human Feces

Bacterial group	Bacterial count (number/g wet weight)
Anaerobes	
Bacteroides	$10^8–10^{10}$
Bifidobacterium	$10^9–10^{10}$
Lactobacillus	$10^3–10^8$
Clostridium	$10^3–10^6$
Fusobacterium	$10^3–10^5$
Eubacterium	$10^8–10^{10}$
Veillonella	<10
Aerobes/facultative aerobes	
Enterobacteria	$10^6–10^9$
Enterococci	$10^5–10^8$
Staphylococci	$<10^3$
Bacillus, Pseudomonas, Proteus, Spirochaetes	$<10^3$

Adapted from Pahren, H. R., *Crit. Rev. Environ. Control*, 417, 187, 1987.

Table 9 Relative Efficacies of Selected Sludge Treatment Methods in Reducing Numbers/Survival Times of Selected Pathogens

Type of process	Effect against pathogens	
	Salmonellae	*Taenia saginata* ova
Anaerobic mesophilic digestion	90% removal 99% with secondary digestion	Infectivity destroyed
Cold digestion, lagooning	90–99% removal for 6–12 months retention: absent from 96% of samples after 2 years	No information: ova are nonviable after 6 months
Aerobic digestion	20–80% reduction	No effect
Lime stabilization	Disinfection if pH $>$ 10: rapid at pH $\geq$11.5	No information: inhibits "hatching" in viability tests
Drying beds	Numbers decline with time: not eliminated completely	No information: ova are nonviable after 6 months
Dewatering with aluminum chlorohydrate	No effect	Presumably no effect

Adapted from Pike, E. B. and Davis, R. D., *Sewage Sludge Stabilisation and Disinfection,* Ellis Horwood, Chichester, England, 1984, chap. 3.

diversity, depending on their presence in the host vector contributing to the sewage and the degree of treatment it has undergone[78] (Table 9). Pahren[79] suggested that high temperatures during initial landfill operations, together with the antibiotic characteristics of leachate would adversely affect the survival of viral and bacterial pathogens in leachate. However, Kinman et al.[80] reported that, although fecal coliforms could not be isolated from 5-year-old lysimeters inoculated with domestic refuse, they could be recovered from lysimeters supplemented with sewage sludge and nutrients. In general, however, the threat posed by the pathogenic content of leachate should not be significant.

**Table 10   Concentrations of
               Selected Heavy Metals in
               Sewage Sludge**

Metal concentration in sludge (g/kg)		
Metal	Raw	Digested
Fe	15.0	20.0
Ni	0.06	0.07
Cr	1.2	1.6
Zn	1.2	1.7
Pb	0.4	0.6

Adapted from Oleszkiewicz, J. A. and
Sharma, V. K., *Biol. Wastes,* 31, 45,
1990.

g. Heavy Metals

Although heavy metal concentrations in sludge remain a matter of concern in the application of sludge to agricultural land, it is unlikely that concentrations of heavy metals present in domestic sludge (Table 10) are high enough to adversely affect leachate quality or refuse decomposition processes.

h. Landfill Gas Production

Landfill gas production has traditionally been regarded as a problem and a liability, due to the risk of explosions and fires, and the necessity to install gas venting systems. Sludge co-disposal with refuse has been shown to accelerate progression toward the methanogenic phase of the refuse fermentation and to increase rates of methane production.[123] Craft and Blakey[123] observed that co-disposal increased rates of gas production, from 0.05 m^3 methane/tonne volatile solids/d (control reactor) to 0.28 to 0.45 m^3/t/d in the test systems.

The effects of added nutrients on methanogenesis in refuse-packed lysimeters were also examined by Leuschner,[81] who operated six lysimeters under varying conditions of buffer, nitrogen (urea), phosphorus, anaerobically digested sludge, and septic tank residue additions. Reactors seeded with both nutrients (N and P) and buffer exhibited methanogenic lag phases 70 days shorter than control reactors. However, continued nutrient addition, following initiation of methanogenesis, did not improve the rate of methane formation compared to buffer-only controls. These data suggested that a differential nutrient demand was exerted on the system during different stages of the fermentation, with a greater requirement during the initial period. Addition of sewage sludge to refuse samples also resulted in a reduced lag phase before the onset of methanogenic activity, which then occurred at a rate significantly greater than in any other reactor. The reduced lag phase was attributed to nutrient supplementation and improved gas release rates to the addition of a viable microbial "seed" population.

i. Benefits of Sludge Co-Disposal

Despite the potential environmental impacts associated with sludge co-disposal, major benefits may be derived from this practice.

Generation of Methane — In the U.K., the benefits, both environmental and energetic, of harnessing landfill gas are increasingly recognized.[80] An emphasis on the development of renewable energy sources and the implementation of the Non-Fossil Fuel Obligation Act (1990), which obliges electricity companies in England and Wales to purchase a proportion of their energy from non-fossil fuel sources, has encouraged the installation of gas collection schemes. Of the first tranche of projects accepted, 25, out of 75, were landfill gas exploitation schemes, with a capacity of $25MW_e$.[80] The economics of such schemes are obviously improved with increased methane production.

Co-disposal of sewage sludge with domestic refuse has been shown to decrease the time period before the onset of methanogenesis. Several authors[81,82] have proposed that co-disposal of sludges offers one of the best prospects for the optimization of methanogenic activity within landfill sites. Sludge addition to refuse is thought to promote methanogenesis by the addition of either nutrients or an appropriate inoculum and moisture, thus facilitating the development of an active methanogenic population.

Increased Stabilization Rate — Several authors[76,123] have observed that sludge co-disposal leads to more rapid landfill stabilization. A variety of benefits may accrue from this which, if achieved, would make sludge co-disposal an attractive practice, both environmentally and economically:

1. Reduction in the period over which a polluting leachate, requiring treatment, is generated.
2. Faster generation of the renewable energy source, methane.
3. A faster "turn over" period from site initiation to ultimate restoration/reclamation, and site license surrender.

Landfill as an Anaerobic Filter — The exploitation of landfill sites as anaerobic filters to attenuate potentially toxic compounds has been promoted,[50,85] and in the majority of co-disposal operations this is probably the primary objective. While disposal of domestic sludges is unlikely to introduce significant quantities of toxic material into the landfill, industrial sludges have the potential to do so.

Table 11 Analysis of H-Coal Effluent

Parameter	Concentration (mg/l)
Organic carbon	7,600
COD	21,100
Phenolics	7,600
Phenol	4,900
o-Cresol	586
m-Cresol	1,230
p-Cresol	420
2,4/2,5-Dimethylphenol	63
3,5-Dimethylphenol	213
3,4-Dimethylphenol	440
Total Kjeldahl nitrogen	267
Nitrite N	0.2
Nitrate N	0.8
Ammonia N	6.4
Total cyanide	0.21
Total phosphorus	5
pH	7.4

Adapted from Fedorak, P. M. and Hrudey, S. E., *Water Sci. Technol.*, 17, 143, 1985.

Economic Factors — Currently, landfill remains a relatively low-cost disposal option compared to other potential outlets for sewage sludge. However, Frost et al.[58] noted that, while current charges for the acquisition of landfill space lie between £0.50 and £1.50 per m^3, the increasing shortage of suitable tipping space, combined with a realization by landowners of their valuable asset, is forcing the price upward.

2. *Industrial Process Effluents*

Organic process effluents are generally characterized by a range of components: H-coal wastewater, for example (Table 11), comprises phenol, a range of substituted phenols, cyanide, and other inorganic species. There are likely, therefore, to be a range of factors, specific to each effluent, mediating the immobilization, dilution, or transformation of the component molecules.

a. *Immobilization*

Immobilization of organic species in refuse is largely due to adsorption. Adsorption may be considered either as a physicochemical process of adhesion from a fluid to solid surfaces or as an immobilization process which confers at least partial resistance to extraction by salt solution.[86]

In soils, organic and inorganic colloid substances such as clay, organic matter, and hydrous oxides of Fe, Al, and Mn form the greater part of the adsorption complex, while the noncolloid materials, such as sand and silt, exhibit adsorption to a much lesser extent. Although adsorption in soils has been extensively studied and reviewed,[86] the extent and significance of adsorptive processes in refuse have received less attention.

The extent and reversibility of adsorption of any specified waste is dependent on a range of factors such as the nature of the waste and the available surfaces, refuse pH, extent of surface area, and anaerobic conditions. The adsorption, for example, of trichloroethylene, a halogenated solvent which is used for decaffeinating coffee, as a dry-cleaning agent, and as a solvent in vapor degreasing, on the representative waste surfaces of unprinted newspaper, 50/50 Dacron/cotton, and polyethylene sheet (1.6 mm thick) has been examined.[87] The adsorption process in the presence of a model aerobic leachate (82 mM acetic acid) buffered at pH 5, 7, or 9 generally approached equilibrium within <5 h, was at least partially reversible, and conformed well with a linear Freundlich isotherm of the form:

$$C_s = K_A C \qquad (1)$$

where C_s is the concentration in the solid substrate, C is the concentration in the leachate, and K_A is a constant. Differential adsorption was observed, with K_A values for polyethylene 2 to 5 times higher than those for textile, and 8 to 20 times higher than those for paper. The nature of the adsorptive surface appeared more influential than the leachate pH as, for example, K_A values for polyethylene increased by a factor <2 between the pH values 5 and 9.

Reinhart et al.[88] characterized the sorptive behavior of a range of organic molecules, such as 1,4-dichlorobenzene, 1,2,4-trichlorobenzene, naphthalene, 2,4-dichlorophenol, and trichloroethane, on different materials characteristic of domestic refuse (Table 12). An increasing affinity was associated with decreasing solid phase surface energy and increasing surface wettability. It was thus concluded that the majority of refuse adsorption would occur on low energy organic surfaces such as fats, oils, microbial cell walls, lignin, plastic, and leather. The authors considered that the retention by adsorption and, consequently, delayed migration of organic molecules enhance the assimilative capacity of a landfill and afford greater opportunity for more complete biological and chemical degradation of the otherwise recalcitrant constituents. Increased recycling of plastics, and reduced plastic content in refuse, may, therefore, influence more than the structural properties of refuse, particularly since image analysis studies have demonstrated significant colonization of plastic materials.[89]

Adsorption of phenol to refuse was investigated by Knox and Newton,[28] who challenged 1-kg batches of both "fresh" (8 weeks) and "aged" (4 years) refuse, under anaerobic conditions, with concentrations of phenol which ranged from 182 to 1817 mg/l. Less than 1 h was required for equilibration, at which point a mean 5.3 or 15% of phenol was adsorbed from

Table 12 Sorption Capacity of Various Refuse Components for Dichlorobenzene, Trichlorobenzene, and Naphthalene

Material	Typical % (w/w) of refuse	1,4-dichlorobenzene		1,2,4-trichlorobenzene		Naphthalene	
		Partition coefficient (l/mg)	Concentration (mg/l)	Partition coefficient (l/mg)	Concentration (mg/l)	Partition coefficient (l/mg)	Concentration (mg/l)
Plastic	5	115	0.33	735	0.42	17	0.61
Wood	2	70	0.39	260	0.89	46	0.10
Paper	40	31	0.44	120	1.10	32	0.65
Tin can	6	20	0.45	39	1.39	2	0.65
Fabric	2	37	0.47	17	1.50	21	0.61
Soil	4	129	0.55	113	0.28	27	0.14

Adapted from Reinhart, D. R., Gould, J. P., Cross, W. H., and Pohland, F. G., *Emerging Technologies in Hazardous Waste Management*, American Chemical Society, Washington, DC, 1989, chap. 17.

solution onto "aged" or "fresh" refuse, respectively. The DOE[19] concluded that adsorption of phenols on "aged" refuse could not be relied upon to reduce the concentrations of phenolic wastes to acceptable levels. Although this study found no evidence of desorption, Knox et al.[90] reported some desorption of phenol and p-cresol, although they were unable to derive meaningful desorption isotherms.

Mobility may also be modified by differential solubilities. Many chemicals are more soluble in organic solvents than in water. Phenol, for example, although soluble in water (82 g/l), is more soluble in a range of organic solvents such as benzene [distribution coefficient (D_c) at 20°C = 2.2], diethylether $(D_c = 17.0)$, iso-propylether $(D_c = 17.0)$ and phenosolvan $(D_c = 59.0)$.[91] Since a wide range of pesticides exhibits similar properties (Table 6), leaching of their wastes should be related partly to the infiltration of aqueous fluids and partly to the presence of organic solvents, such as waste oils. Thus, the occurrence of heptachlor and aldrin in the leachates of several hazardous waste sites[52] may reflect both their biological recalcitrance and their mobilization by organic solvent-containing wastes. The continuous leaching of concentrations, however low, of molecules such as aldrin cannot be regarded as an attenuating mechanism since significant bioaccumulation has been demonstrated in algae (*Chlorella fusca*), fish (Golden Ide), and rats.[92]

b. Dilution

Dilution of a waste by the infiltration of liquid, whether rainfall or wastewater, to a landfill site, and the dispersion of the leachate in the unsaturated zone, may reduce significantly the concentration of the xenobiotic. Although Feates and Parker[8] considered that "so-called toxic materials" could be diluted to a concentration "below the hazard level," it has been observed[93] that the carcinogenicity of many chemicals has not been tested and that epidemiological data may take 60 years to accumulate. Moreover, bioaccumulation data for many chemicals are not available, although Freitag et al.[92] examined bioaccumulation of a range of organic chemicals in the freshwater fish, the Golden Ide, and observed, over 3 days, accumulation factors of 3850, 2760, 1900, 260, and 20 for 2,5,4-trichlorobiphenyl, aldrin, DDT, pentachlorophenol, and phenol, respectively. The bioaccumulation of ions by both animals and plants, with subsequent accumulation in man and the possible development of pathological conditions, has been reviewed.[3]

Unfortunately, dilution may also militate against biodegradation. Since Boethling and Alexander[37] first reported that the percentage mineralization of 2,4-D at a concentration of 2.2 μg/l, was only 1/16th that at 220 μg/l, there have been further observations that, for some chemicals, such as 1-naphthyl-N-carbamate and phthalate esters,[94] a threshold concentration exists below which rates of mineralization are no longer linearly related to concentration. Further anomalies in the transformation of low concentrations of organic compounds have since been observed.[95] Mineralization may, for example, proceed in a biphasic manner, and acclimation to a substrate may occur even though the substrate concentration is below the threshold value.

It seems likely that threshold concentrations will vary with the microorganism or population, although no reports have, as yet, implicated anaerobic microbial associations. Lack of knowledge of the effects of low substrate concentrations on growth, enzyme induction, and enzyme activity and lack of data on specific chemicals make it difficult to assess the *in situ* significance of threshold concentrations. These may, however, account for the persistence of low concentrations of biodegradable organic substances in natural environments.[94]

Possible environmental and health effects and limited biodegradation of low concentrations of diluted waste chemicals may thus limit the efficacy of the "dilute-and-disperse" mechanism. The method suffers additionally from lack of control, the necessity for the production of large volumes of leachate, and from a negative image in the eyes of the public.

c. Transformation

Volatilization and transfer of the pollution potential of the wastewater into the atmosphere have been demonstrated for several pollutants such as dichloromethane, carbon tetrachloride, benzene, *ortho-*, *meta-*, and *para*-cresols, chlorobenzenes, and trichloroethane.[96] Not only will volatilization, and the resultant odors, adversely invoke public opinion but regulatory atmospheric limits may be exceeded. Assmuth and Kalevi,[96] for example, demonstrated that, at selected Finnish landfills, occupational health norms were occasionally exceeded by carbon tetrachloride in particular.

The most effective attenuation process is the biological transformation of an organic waste, with resultant complete mineralization or partial degradation to a nontoxic and, perhaps, further degradable intermediate. The unrecognized potential for organic chemical degradation of anaerobic or anaerobic/aerobic interfacial systems, such as landfill, was emphasized by Willson et al.[97] when they observed that the initial step in the biodegradation of DDT was reductive and occurred readily under soil anaerobiosis. Degradation under anaerobic conditions may, however, favor the production of toxic metabolites. Gandolla and Aragno[98] considered that biotests were essential to detect conversions such as the sequential degradation under methanogenic conditions of tetrachloroethylene, a widely used compound of low toxicity, to tri- and dichlorinated compounds, and finally to the highly toxic vinyl chloride. A range of such test protocols has been described,[99,100] although their limitations should be recognized. The bio-chemical methane potential (BMP) assay[100] is operationally simple and convenient; however, the addition of the molecule to fresh refuse may not adequately reflect the potential impact of the xenobiotic on active refuse fermentations. It has been shown, for example, that, although the addition of 7 mM o-cresol to fresh refuse inhibits the development of a methanogenic fermentation, addition to a methanogenically active refuse sample has no effect on methane release.[101] The method of Benstead et al.,[99] which is dependent of membrane inlet mass spectrometry to detect shifts in dissolved methane and hydrogen concentrations, and uses a co-culture of *Desulfovibrio desulfuricans* and *Methanobacterium bryantii*, is primarily a research tool.

d. Biodegradation of Organic Wastes

The anaerobic biodegradation in refuse of specific organic wastes has, as yet, only been critically assessed for phenolic wastes, although there is increasing interest in the co-disposal of vegetation wastewaters[102] and the degradation of chlorinated aliphatic compounds (1,1,1-trichloroethane, te-trachloromethane, and tetrachloroethylene) was recorded both in laboratory reactors containing landfill leachate and in landfill *in situ* simulators.[122]

In a comprehensive series of experiments[103–105] refuse columns were used to examine the effects of both phenol loading and irrigation rates, and leachate perfusion strategies on the refuse fermentation. The co-disposal of low (1 mM) phenol concentrations partially inhibited the development of methanogenesis in columns operated with leachate discard,[103] although with leachate recycle inhibition was only recorded at concentrations ≥ 4 mM. Leachate recycle also proved an efficacious strategy to promote biod-egradation of phenol: at a dilution rate of 0.03/h, and over 524 days, 100% removal of influent 2 mM phenol was recorded, compared to 31.5% removal of influent phenol from columns operated with leachate discard. Although methane release rates were variable, there were indications that the co-disposal of phenol and subsequent biodegradation of the molecule resulted in increased methane release rates.

Methane release rates were observed to increase with phenol concen-trations in a multistage refuse column array, perfused with a range (2 to 6 mM) of phenol concentrations (Figure 2).[104] At a dilution rate of 0.07/h, 100% removal of ≤ 6 mM phenol was recorded, although perfusion of 8 mM phenol resulted in progressive inhibition of phenol dissimilation, and reduced methane release, although simultaneous dosing with hexanoic acid (5 mM), a characteristic leachate component,[45] showed that the β-oxidation pathway remained fully operational with no inhibition of methanogenesis, suggesting that the phenol degraders were more susceptible to the effects of phenol than the methanogens. This interpretation was supported by the

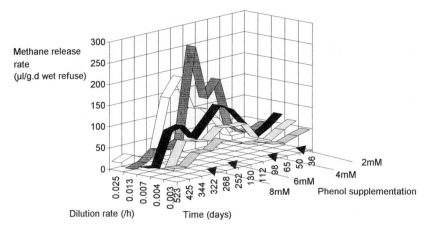

Figure 2   Changes in methane release rates in a multi-stage refuse column array, at
sampling ports corresponding to empty-bed dilution rates of 0.025, 0.013,
0.007, 0.004, and 0.003/h, with increasing influent phenol supplementation
(2 to 8 mM). (Adapted from Watson-Craik, I. A. and Senior, E., *J. Chem.
Technol. Biotechnol.*, 47, 219, 1990.)

results of a separate study which examined the effects of phenol wastewater
co-disposal on the anaerobic biodegradation of hexanoic acid.[106] By use of
a multi-stage continuous culture model, sulfate-reducing bacteria and methanogens were shown to be inhibited by phenol concentrations ≥2 and ≥8
mM, respectively (Figure 3). With 8 mM phenol, there was no evidence of
inhibition of β-oxidation.

The anaerobic degradation of phenol in both anaerobic digesters and
refuse columns was reviewed by Knox and Gronow,[85] who concluded that
the existence of one or more of saturated zone/leachate recycle, established
methanogenesis, and elevated temperatures was likely to enhance rates of
phenol removal, which were suggested could range from 5 to 50 g/m³/d.

D. INORGANIC WASTES

The effective co-disposal of substances which are either inherently
nonbiodegradable, such as heavy metals, or relatively nonbiodegradable,
such as some pesticide residues, is dependent on dilution to environmentally acceptable concentrations, or to immobilization of the waste in the
landfill site. Immobilization may be effected either by physico-chemical
factors which stabilize a waste at a fixed locus or by provision of a site
liner. In the latter case, the provision of leachate collection and treatment
should also be considered.

Although effective immobilization necessitates an understanding of the
possible fixation mechanisms and of their relative significance for specific

(a)

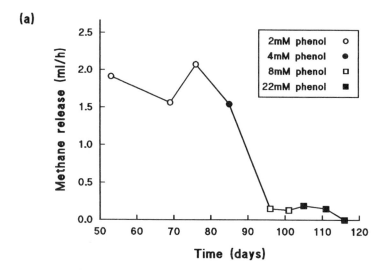

(b)

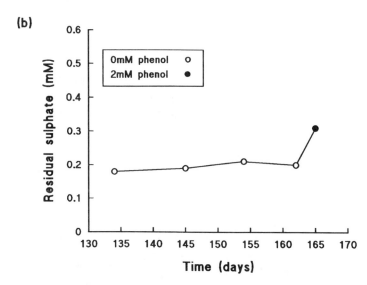

Figure 3 Changes in (a) methane release rates in the third vessel and (b) residual sulfate concentrations in the first vessel of a three-stage continuous culture array, when the influent to the respective vessel was supplemented with incrementally increasing (2 to 22 m*M*) phenol concentrations. (Adapted from Watson-Craik, I. A., and Senior, E., *Lett. Appl. Microbiol.*, 9, 227, 1989.)

wastes, these, and in particular those mediated by biological activity, have been subjected to only limited critical assessment.

1. Refuse Mass Absorption

Immobilization of a liquid waste by absorption into the refuse mass may only be achieved through an understanding and manipulation of the site water balance. The field capacity[107] may be defined as the maximum amount of liquid that the refuse mass can hold before infiltrating liquid appears as leachate. However, since emplaced refuse has an inherent moisture content, the absorptive capacity, i.e., is the volume of additional liquid which can be taken up prior to leachate generation, may be of more relevance.

The complex of factors, which includes ground- and surface-water infiltration, annual precipitation, evaporation, infiltration capacity, moisture content, composition, and density of emplaced refuse, which influences the absorptive capacity of a site, has been extensively reviewed.[30,31,107,108]

The practical difficulties involved in the assessment *in situ* of the absorptive capacity of a site result, in many cases, in the discharge of liquid until the work area becomes saturated. The DOE[19] noted that in two sites problems had resulted from the breakout of liquid waste upon compaction and that at another site the discharge of sludge together with liquid waste had led to the formation of a lagoon and, subsequently, to an overspill. It is clear that absorption cannot be relied on for the long-term immobilization of waste materials.

2. Refuse/Soil Surface Adsorption

Few studies have addressed the elucidation of influential parameters on the adsorption of inorganic wastes, such as heavy metals, on refuse surfaces, although the adsorption of both arsenic (III) and (V) by "aged" and "fresh" domestic refuse was examined by Blakey[33] who observed that, over the concentration range 1.25 to 100 mg arsenic III/l leachate, concentrations were reduced by 96% at pH 9 and 57% at pH 5, in "aged" refuse. In "fresh" domestic refuse, the equivalent reductions were 86 and 43%, respectively. Adsorption of arsenic was thus reduced at low leachate pH and was more significant in "aged" refuse. A comparison, however, of both the methods and results of this study with those of Jones et al.[87] exemplifies the inherent practical difficulties confronting the potential co-disposal practitioner. Not only does the heterogeneity of received refuse militate against an accurate estimation of the significance of individual attenuating mechanisms, but also the absence of standard procedures for the determination of adsorption characteristics of specific wastes, on either specified refuse surfaces or characterized refuse, may result in conflicting data, and thus the adoption of site strategies whose potential environmental impact is unrecognized.

Quantitative data on the desorption characteristics of domestic refuse are also lacking. Thus, for example, although Blakey[33] recorded <5% desorption of As(V) at pH values of 5 or 7, Jones et al.[87] reported the "partial" desorption of cadmium, arsenic, or trichloroethylene. As with adsorption studies, the lack of standardization of experimental procedures militates against the acquisition of comparable data. Although extraction with water[109] may be expected to yield water-soluble, very loosely retained trace metals, extraction with an acid solution[87,110] may reflect more accurately the conditions which pertain within a landfill site.

3. pH Effects

The degree of leaching of metals from, for example, metal hydroxide sludges may also be influenced by leachate pH. Thus, the leachability of lead has been shown to increase exponentially with decreasing pH, while cadmium leachability also increased.[111]

In general, the solubility of heavy metals and, indeed, many organic pollutants, is higher at pH <4 than at pH >7,[112] with implications not only for the disposal of acid wastes but for sites where a balanced methanogenic fermentation is not operative. Leachate pH values as low as 5.12 have been recorded in young landfill sites.[121]

4. Complexation/Chelation Reactions

With the identification of significant concentrations of humic and fulvic acids in landfill leachates,[113] the argument for organic complexation as an influential factor in pollutant solubility counteracting attenuation, particularly of metals, was presented. The carboxylic (-COOH) and phenolic (-OH) groups of humic and fulvic acids are highly reactive and cations, minerals, and certain organic constituents may be bound to the acids through these groups. In addition, complexation with the reactive groups of low molecular weight compounds such as volatile fatty acids (acetic, butyric, propionic), aromatic acids (such as the lignin decomposition products, vanillic, syringic, and protocatechuic acids), amino acids (aspartic, glutamic), and phenol may further enhance the downward movement of metals as soluble metal-organic matter complexes. Gould et al.[112] considered that chloride was also an important complexing agent.

The significance of complexation as a mobilization mechanism in co-disposal sites is uncertain, although Blakey[114] reported a significant correlation in a range of landfill leachates between zinc, nickel, lead, and total organic carbon (TOC) concentrations. A similar correlation between leachate zinc and aromatic hydroxyl concentrations was also observed by Pohland et al.[115]

5. Precipitation

Hydrogen sulfide, produced by sulfate-reducing bacterial activity, is a strong reducing agent and can mediate the precipitation of metals such as iron, copper, and mercury by reducing them to their insoluble sulfides. Sulfides are highly insoluble, with solubility products (K) which range from 1×10^{-48} (Cu_2S) to 1×10^{-18} (FeS).[116] Gould et al.[112] reported that chromium, unlike other metals such as zinc, nickel, and cadmium, was precipitated as the insoluble hydroxide at the oxidation-reduction potential values characteristic of operative landfills (-50 to -500 mV).

Although Mosey et al.[116] suggested that, in anaerobic digesters, phosphates could also reduce heavy metal toxicity since, for example, the solubility product of $PbPO_4$ is 3×10^{-44}, the low concentrations of phosphates generally present in leachates[48,66] suggest that this process is unlikely to be significant in landfills.

Train[18] reported that added protection from leaching was often provided by the chemistry of calcium or magnesium carbonates, since the presence of any limestone or dolomite would ensure an excess of soluble hydroxy-carbonates to precipitate soluble ions. Carbonate solubility products range from 6×10^{-11} (Zn) to 1.5×10^{-13} (Pb).[116] Thus, while hard-water areas provide a degree of *in situ* protection, soft-water areas may pose greater problems.

6. Reducing Conditions

In addition to precipitation by H_2S and complexation with organic acids, reducing conditions, created by biological activity, may have further repercussions on attenuation. The heavy metals Ni, Cr, Zn, Pb, Co, Fe, and Mn, for example, are more mobile under anaerobic conditions, all other factors being the same.[84,117] Phosphates may be released from ferric ores by the activities of both sulfate-reducing[118] and iron-reducing[119] bacteria. However, since phosphate is often considered to be present in limiting concentrations in refuse,[41] it may be speculated that the released phosphates are unlikely to pose an environmental challenge.

Of the mechanisms described above, it has been suggested[120] that sulfide and carbonate precipitation, rather than adsorption on refuse components, may provide the major controls over metal mobility from landfills. If this is so, Gould et al.[112] observed that immobilization of metals is reliant more on the biochemical regime within the landfill site than on the nature of the refuse itself.

E. CO-DISPOSAL IMPLEMENTATION

All recommendations/practices/strategies must be capable of implementation on site. Thus, for example, since facilities on site for the rapid analyses of wastes are not practicable, waste loads must be accurately labeled. At present, site licenses may define the maximum permitted concentration of a waste, but there is no practical method of determining the actual concentrations at the site.

Since, at present, official guidelines on operational parameters, such as loading rates and permissible waste mixtures, are lacking, the onus is on the site manager to exercise expertise in deciding the type and quantities of waste suitable for the site. However, in spite of the lack of guidelines, the site license holder is still legally responsible for any resultant environmental hazard or pollution. Realistic guidelines and strategies will only result from an understanding of the fundamental microbiology and biochemistry of co-disposal, derived from environmentally relevant model systems.

REFERENCES

1. Cope, C. B., Hazardous wastes and their recycling potential, in *The Scientific Management of Hazardous Wastes,* Cope, C. B., Fuller, W. H., and Willetts, S. L., Eds., Cambridge University Press, Cambridge, 1983, chap. 1.
2. Hazardous Waste and Consolidated Permit Regulations, Vol. 45, No. 98, Book 2, U.S. Environmental Protection Agency, Washington, D.C., 1980.
3. The Licensing of Waste Disposal Sites, WMP No. 4, Department of the Environment, Her Majesty's Stationery Office, London, 1976.
4. Sands, P., Emerging trends in EC regulation of chemical waste, *Chem. Ind.,* 19, 734, 1992.
5. Brand, E. C., Hazardous waste management in the European Community. Implications of ''1992'', *Sci. Total Environ.,* 129, 241, 1993.
6. Laurence, D. and Wynne, B., Transporting waste in the European Community. A free market?, *Environment,* 31, 12, 1989.
7. Cook, J. D., Landfill as a disposal route for difficult wastes, *Chem. Ind.,* 17, 615, 1984.
8. Feates, F. S. and Parker, A., Codes of practice and research relating to landfill disposal of hazardous waste in the UK, in *Toxic and Hazardous Waste Disposal,* Vol. 2, Pojasek, R. B., Ed., Ann Arbor Science, Ann Arbor, MI, 1979, 51.
9. MacNeill, J. W., Policy issues concerning transfrontier movements of hazardous waste, in *Transfrontier Movements of Hazardous Wastes,* Organisation for Economic Cooperation and Development, Paris, 1985.

10. Department of the Environment, Waste and recycling, in *Digest of Environmental Protection and Water Statistics,* Her Majesty's Stationery Office, London, 1990, chap. 5.

11. Tucker, S. P. and Carson, G. A., Deactivation of hazardous chemical wastes, *Environ. Sci. Technol.,* 19, 215, 1985.

12. Reed, S. B., Monitoring and long-term planning: essential ingredients of waste management, *Conserv. Recyc.,* 7, 107, 1984.

13. Hazardous Waste Disposal, Vol. 1, House of Lords Select Committee on Science and Technology, Her Majesty's Stationery Office, London, 1981.

14. Pearce, P., Landfilling: a long-term option for hazardous waste disposal?, *UNEP Ind. Environ. Spec. Issue,* 1983, p. 57.

15. Hazardous Waste Management: "Ramshackle and Antediluvian?", 2nd Rep., Hazardous Waste Inspectorate, Department of the Environment, London, 1986.

16. Worthy, W., Hazardous waste: treatment technology grows, *Chem. Eng. News,* 60, 10, 1982.

17. Lederman, P. B. and LaGrega, M. D., Managing hazardous wastes, *Ind. Wastes,* 27, 36, 1981.

18. Train D., Treatment and disposal of toxic wastes, *Chem. Eng.,* 390, 35, 1983.

19. Cooperative Programme of Research on the Behaviour of Hazardous Wastes in Landfill Sites, Final Rep., Department of the Environment, Her Majesty's Stationery Office, London, 1978.

20. Paigen, B., Goldman, L., Highland, J. H., Magnant, M. M., and Steegman, A. T., Prevalence of health problems of children living near Love Canal, *Haz. Wastes Haz. Mater.,* 2, 23, 1985.

21. Goldman, L., Paigen, B., Magnant, M. M., and Highland, J. H., Low birth weight, prematurity and birth defects in children living near the hazardous waste site, Love Canal, *Haz. Wastes Haz. Mater.,* 2, 209, 1985.

22. Kennedy, K. J. and van den Berg, L., Anaerobic downflow stationary fixed film reactors, in *Comprehensive Biotechnology,* Vol. 4, Robinson, C. W. and Howell, J. A., Eds., Pergamon Press, Oxford, 1985, 1027.

23. Spain, J. C. and van Veld, P. A., Adaptation of natural microbial communities to degradation of xenobiotic compounds: effects of concentration, exposure time, inoculum and chemical structure, *Appl. Environ. Microbiol.,* 35, 428, 1983.

24. Harris, R. C., Landfill leachate generation—a water authority viewpoint, *Wastes Manage.,* 73, 606, 1983.

25. Rees, J. F., Optimisation of methane production and refuse decomposition in landfills by temperature control, *J. Chem. Technol. Biotechnol.,* 5, 30, 458, 1980.

26. Rees, J. F., Landfills for treatment of solid wastes, in *Comprehensive Biotechnology,* Vol. 4, Robinson, C. W. and Howell, J. A., Eds., Pergamon Press, Oxford, 1985, 1071.

27. Kasali, G. B., Solid-state refuse methanogenic fermentation: control and promotion by water addition, *Lett. Appl. Microbiol.,* 11, 22, 1990.

28. Knox, K. and Newton, J. R., Study of Landfill Disposal of Acid Tars and Phenol-Bearing Lime Sludges. I. Characteristics of Water-soluble Constituents of Wastes and Preliminary Study of Interaction of Phenolic Solutions with Domestic Refuse, WLR Tech. Note Ser. No. 19, Department of the Environment, London, 1976.

29. Newton, J. R., Pilot-scale studies of the leaching of industrial wastes in simulated landfills, *Water Pollut. Control,* 76, 468, 1977.

30. Marriott, J., Some practical aspects of control of leachate from landfill sites, *Solid Wastes,* 71, 513, 1981.

31. Holmes, R., The water balance method of estimating leachate production from landfill sites, *Solid Wastes,* 70, 20, 1980.

32. Khan, A. Q., Co-disposal of wastes—a positive view of sanitary landfill by a Pollution Control Officer, in *Practical Waste Management,* Holmes, J. R., Ed., John Wiley & Sons, Chichester, England, 1983, chap. 10.

33. Blakey, N. C., Behaviour of arsenical wastes co-disposed with domestic solid wastes, *J. Water Pollut. Control Fed.,* 56, 69, 1984.

34. Wood Preserving Wastes, WMP No. 16, Department of the Environment, Her Majesty's Stationery Office, London, 1980.

35. Cheyney, A. C., Experience with the co-disposal of hazardous wastes with domestic refuse, *Chem. Ind.,* 17, 609, 1984.

36. Lee, G. F., Jones, A., and Ray, C., Sanitary landfill leachate recycle, *BioCycle,* 27, 36, 1986.

37. Boethling, R. S. and Alexander, M., Effects of concentration of organic chemicals on their biodegradation by natural microbial communities, *Appl. Environ. Microbiol.,* 37, 1211, 1979.

38. Pahm, M. A. and Alexander, M., Selecting inocula for the biodegradation of organic compounds at low concentrations, *Microb. Ecol.,* 25, 275, 1993.

39. Rittmann, B. E., Microbiological detoxification of hazardous organic contaminants: the crucial role of substrate interaction, *Water Sci. Technol.,* 25, 403, 1992.

40. Saez, P. B., Quantification of Interactions between Inhibitory Primary and Secondary Substrates, Ph.D. Thesis, University of Illinois, Urbana, IL, 1990.

41. Senior, E., Introduction, in *Microbiology of Landfill Sites,* Senior, E., Ed., CRC Press, Boca Raton, FL, 1990, 2.

42. WMP Landfill Practices Sub-Group 2, Section on Loading Rates, Rep., Department of the Environment, London, 1982.

43. Newton, J. R., Pilot-scale Experiments on Leaching from Landfills. II. Experimental Procedures and Comparison of Overall Performance of All Experimental Units. Nov. 1973–Aug. 1975, WLR Tech. Note Ser. No. 17, Department of the Environment, London, 1976.

44. Robinson, H. D., and Maris, P. J., The treatment of leachates from domestic waste in landfill sites, *J. Water Pollut. Control Fed.,* 57, 30, 1985.

45. Harmsen, J., Identification of organic compounds in leachate from a waste tip, *Water Res.,* 17, 699, 1983.

46. Robinson, H. D., Barber, C., and Maris, P. J., Generation and treatment of leachate from domestic wastes in landfills, *Water Pollut. Control,* 81, 465, 1982.

47. Ventakaramani, E. S., Ahlert, R. C., and Corbo, P., Biological treatment of landfill leachates, *Crit. Rev. Environ. Control,* 14, 333, 1984.
48. Robinson, H. D. and Maris, P. J., Leachates from Domestic Wastes: Generation, Composition and Treatment; a Review, Tech. Rep. TR 108, Water Research Centre, Stevenage, U.K., 1979.
49. Crutcher, A. J. and Rovers, F. A., Temperature as an indicator of landfill behaviour, *Water Air Soil Pollut.,* 17, 213, 1982.
50. Watson-Craik, I. A., Landfill as an Anaerobic Filter for the Co-Disposal of Phenolic Wastewaters, Ph.D. Thesis, University of Strathclyde, Glasgow, Scotland, 1987.
51. Hazardous Waste Management: an Overview, 1st Rep., Hazardous Waste Inspectorate, Department of the Environment, London, 1985.
52. Knox, K., Leachate production, control and treatment, in *Hazardous Waste Management Handbook,* Porteous, A., Ed., Butterworths, London, 1985, chap. 4.
53. Westlake, K., Sayce, M., and Fawcett, T., Environmental Impacts from Landfills Accepting Non-Domestic Wastes, Rep. No. CWM/036/91, Department of the Environment, London, 1991.
54. Horan, N. J., *Biological Wastewater Treatment Systems,* John Wiley & Sons, Chichester, England, 1990, chap. 1.
55. Lowe, P., Developments in sewage sludge incursation, in *Where Will all the Sludges Go?* Paper presented at a meeting of the Water & Environmental Group of the SCI, London, 1987.
56. Baker, G., The burning question of sewage treatment, *World Water,* 12, 24, 1989.
57. Anon., Cleaning up the Sewage Business, ENDS Report No. 182, 12, 1990.
58. Frost, R., Powlesland, C., Hall, J. E., Nixon, S. C., and Young, C. P., Review of Sludge Treatment and Disposal Techniques, WRC Report No: PRD 2306-M/1, Stevenage, U.K., 1990.
59. Davis, J. and Garvey, D., Sludge disposal thinking similar in UK, US, *Water Eng. Manage.,* 133, 25, 1986.
60. Garrett, P., A directive for all waters, *Water Bull.,* No. 360, 6, 1990.
61. Bruce, A. M. and Davis, R. D., Sewage sludge disposal: current and future options, *Water Sci. Technol.,* 21, 1113, 1989.
62. Isaac, P. C. G., Sludge disposal—look into the future, *Wastes Manage. Bull.,* 10, 8, 1985.
63. Kelly, J., Landfill option fades, *Water Bull.,* 463, 6, 1991.
64. Oake, R., Sludge disposal, *Inst. Water Offices J.,* 27, 9, 1991.
65. Brunner, P. H. and Lichtensteiger, T., Landfilling of sewage and sludge—practice and legislation in Europe, in *Treatment of Sewage Sludge: Thermophilic Aerobic Digestion and Processing Requirements for Landfilling,* Bruce, A. M., Colin, F., and Newman, P. J., Eds., Elsevier, London, 1989.
66. Cossu, R., Blakey, N. C., and Traponi, P., Degradation of mixed solid wastes in conditions of moisture saturation, in *Proc. Int. Symp. Process Technol. Environ. Impact Sanitary Landfill,* Vol. 1, CISA, Cagliari, Italy, October 19–23, 1987.
67. Coutts, D. A. P., Dunk, M., and Pugh, S. Y. R., The microbiology of landfills—the Brogborough landfill test cells, in *Landfill Microbiology: R&D Workshop,* Lawson, P. S. and Alston, Y. R., Eds., Harwell Laboratories, Harwell, U.K., 1990, 133.

68. Harries, C. R., Landfill microbiology—work supported at Biotal by the Department of Environment, in *Landfill Microbiology: R&D Workshop,* Lawson, P. S. and Alston, Y. R., Harwell Laboratories, Harwell, U.K., 1990, 150.

69. Scott, M. P., Options for the Treatment of Municipal and Chemical Waste Leachate, paper presented at Conference on Hazardous Waste Handling and Disposal, Penticton, British Columbia, 1981.

70. Robinson, H. D. and Maris, P. J., Leachate from Domestic Waste: Generation Composition and Treatment: A Review, Water Research Centre, Technical Report TR 108, Stevenage, U.K., March 1979.

71. Senior, E., Introduction, in *Microbiology of Landfill Sites,* Senior, E., Ed., CRC Press, Boca Raton, FL, 1990, chap. 1.

72. Watson-Craik, I. A., Sinclair, K. J., and Senior, E., Landfill co-disposal of wastewaters and sludges, in *Microbial Control of Pollution,* Fry, J. C., Gadd, G. M., Herbert, R. A., Jones, C. W., and Watson-Craik, I. A., Eds., Cambridge University Press, Cambridge, 1992, 129.

73. Barlaz, M. A., Schaefer, D. M., and Ham, R. K., Effects of prechilling 2nd sequential washing on enumeration of microorganisms from refuse, *Appl. Environ. Microbiol.,* 55, 50, 1989.

74. Department of Energy, A Basic Study of Landfill Microbiology and Biochemistry, Contractor's Report, AFRC Institute of Food Research, ETSU B 159, London, 1988.

75. Sinclair, K. J., The Co-Disposal of Sewage Sludge with Domestic Refuse and Potential Importance of Landfill Nitrogen Transformations, Ph.D. Thesis, University of Strathclyde, Glasgow, Scotland, 1994.

76. Buivid, M. G., Wise, D. L., Blanchet, M. J., Remedios, E. C., Jenkins, B. M., Boyd, W. F., and Pacey, J. G., Fuel gas enhancement by controlled landfilling of municipal solid waste, *Resourc. Conserv.,* 6, 3, 1981.

77. Barlaz, M. A., Ham, R. K., and Schaefer, D. M., Methane production from municipal refuse: a review of enhancement techniques and microbial dynamics, *Crit. Rev. Environ. Control,* 19, 557, 1990.

78. Winkler, M., Sewage sludge treatments, *Chem. Ind.,* 7, 327, 1993.

79. Pahren, H. R., Microorganisms in municipal solid waste and public health implications, *CRC Crit. Rev. Environ. Control,* 17, 187, 1987.

80. Kinman, R., Rickabough, J., Nutini, D., and Lambert, M., Gas Characterization, Microbial Analysis and Disposal of Refuse in GRI Landfill Simulators, EPA-600/2-86-041 Hazardous Waste Engineering Research Laboratory, Cincinnati, OH, 1986.

81. Leuschner, A. P., Enhancement of degradation: laboratory scale experiments, in *Sanitary Landfilling: Process Technology and Environmental Impact,* Christensen, T. H., Cossu, R., and Stegmann, R., Eds., Academic Press, London, 1989, 83.

82. Evans, S. A., The role of microbiology in harnessing landfill gas as an alternative energy resource, in *Landfill Microbiology: R & D Workshop II,* Evans, S. A. and Lawson, P. S., Eds., Harwell Laboratories, Harwell, U.K., 1991, 21.

83. Anon., UK Ministers in the dark, *Chem. Ind.,* 26, 370, 1991.

84. Gendebien, A. and Nyns, E. J., Biotechnology of sanitary landfilling, presented at the Int. Symp. Environ. Biotechnol., April 22–25, 1991, Oostende, Belgium, 66th event of the European Federation of Biotechnology. Part I, 165, 1991.

85. Knox, K. and Gronow, J., A reactor based assessment of co-disposal, *Waste Manage. Res.,* 8, 255, 1991.

86. Fuller, W. H., The geochemistry of hazardous waste material, in *The Scientific Management of Hazardous Wastes,* Cope, C. B., Fuller, W. H., and Willetts, S. L., Eds., Cambridge University Press, Cambridge, 1983, chap. 8.

87. Jones, C. J., McGuigan, P. J., Smith, A. J., and Wright, S. J., Adsorption of some toxic substances by waste components, *J. Haz. Mater.,* 2, 219, 1978.

88. Reinhart, D. R., Gould, J. P., Cross, W. H., and Pohland, F. G., Sorptive behaviour of selected organic pollutants codisposed in a municipal landfill, in *Emerging Technologies in Hazardous Waste Management,* Tedder, D. W. and Pohland, F. G., Eds., American Chemical Society, Washington, DC, 1990, chap. 17.

89. Jones, R. J., Watson-Craik, I. A., and Senior, E., Image analysis of the effect of different materials on surface colonisation by anaerobic microbial associations from landfill sites, *Binary,* 6, 78, 1994.

90. Knox, K., Newton, J. R., and Stiff, M. J., Study of Landfill Co-disposal of Acid Tars and Phenol-bearing Lime Sludges. II. Mathematical Model and Further Laboratory Studies, WLR Tech. Note Ser. No. 52, Department of the Environment, London, 1977.

91. Verschueren, K., *Handbook of Environmental Data on Organic Chemicals,* Van Nostrand Reinhold, New York, 1977.

92. Freitag, D., Ballhorn, L., Geyer, H., and Korte, F., Environmental hazard profile of organic chemicals: an experimental ecosphere by means of simple tests with C-14 labelled chemicals, *Chemosphere,* 14, 1589, 1985.

93. Hileman, B., Water quality uncertainties, *Environ. Sci. Technol.,* 18, 124A, 1984.

94. Alexander, M., Biodegradation of organic chemicals, *Environ. Sci. Technol.,* 18, 106, 1985.

95. Hoover, D. G., Borgonovi, G. E., Jones, S. H., and Alexander, M., Anomalies in mineralization of low concentrations of organic chemicals in lake water and sewage, *Appl. Environ. Microbiol.,* 51, 226, 1986.

96. Assmuth, T. and Kalevi, K., Concentrations and toxicological significance of trace organic compounds in municipal solid-waste landfill gas, *Chemosphere,* 24, 1207, 1992.

97. Willson, G. B., Parr, J. F., Taylor, J. M., and Sikora, L. J., Land treatment of industrial wastes; principles and practices, *BioCycle,* 23, 37, 1982.

98. Gandolla, M. and Aragno, M., The importance of microbiology in waste management, *Experientia,* 48, 362, 1992.

99. Benstead, J., Archer, D. B., and Lloyd, D., Rapid method for monitoring methanogenic activities in mixed culture: effects of inhibitory compounds, *Biotechnol. Techniques,* 7, 31, 1993.

100. Bogner, J. E., Controlled study of landfill biodegradation rates using modified BMP assays, *Waste Manage. Res.,* 8, 329, 1990.

101. Watson-Craik, I. A., Sinclair, K. J., James, A. G., Sulisti, and Senior, E., Studies of the refuse methanogenic fermentation by use of laboratory models, *Water Sci. Technol.,* 27, 15, 1993.

102. Cossu, R., Blakey, N. C., and Cannas, P., Influence of codisposal of municipal solid waste and olive vegetation water on the anaerobic digestion of a sanitary landfill, *Water Sci. Technol.,* 27, 261, 1993.

103. Watson-Craik, I. A. and Senior, E., Treatment of phenolic wastewaters by co-disposal with refuse, *Water Res.,* 23, 1293, 1989.

104. Watson-Craik, I. A. and Senior, E., Landfill co-disposal of phenol-bearing wastewaters: organic load considerations, *J. Chem. Technol. Biotechnol.,* 47, 219, 1990.

105. Watson-Craik, I. A. and Senior, E., Landfill co-disposal: hydraulic loading rate considerations, *J. Chem. Technol. Biotechnol.,* 45, 203, 1989.

106. Watson-Craik, I. A. and Senior, E., Effects of phenol wastewater co-disposal on the attenuation of the refuse leachate molecule hexanoic acid, *Lett. Appl. Microbiol.,* 9, 227, 1989.

107. Campbell, D. J. V., Understanding water balance in landfill sites, *Wastes Manage.,* 73, 59, 1983.

108. Blakey, N. C., Infiltration and absorption of water by domestic wastes in landfills—research carried out by the Water Research Centre, in *Landfill Leachate Symp. Papers,* paper 4, Harwell Laboratories, Harwell, U.K., 1982.

109. Bridle, T. R., Côté, P. L., Constable, T. W., and Fraser, J. L., Evaluation of heavy metal leachability from solid wastes, *Water Sci. Technol.,* 19, 1029, 1987.

110. Warner, J. S., Hidy, B. J., Jungclaus, G. A., McKown, M. M., Miller, M. P., and Riggin, R. M., Development of a method for determining the leachability of organic compounds from solid wastes, in *Hazardous Solid Waste Testing: First Conf.,* ASTM Spec. Tech. Publ. 760, Conway, R. A. and Malloy, B. C., Eds., American Society for Testing and Materials, Philadelphia, PA, 1981, 40.

111. Josephson, J., Immobilization and leachability of hazardous wastes, *Environ. Sci. Technol.,* 16, 219A, 1982.

112. Gould, J. P., Cross, W. H., and Pohland, F. G., Factors influencing mobility of toxic metals in landfills operated with leachate recycle, in *Emerging Technologies in Hazardous Waste Management,* Tedder, D. W. and Pohland, F. G., Eds., American Chemical Society, Washington, DC, 1990, chap. 16.

113. Artiola-Fortuny, J. and Fuller, W. H., Phenols in municipal solid waste leachates and their attenuation by clay soils, *Soil Sci.,* 133, 218, 1982.

114. Blakey, N. C., Co-disposal of hazardous waste in domestic waste landfills, in Proc. ISWA, Environ. 1988, Amsterdam, Sept. 1988.

115. Pohland, F. G., Gould, J. P., and Ghosh, S. B., Management of hazardous wastes by landfill codisposal with municipal refuse, *Haz. Waste Haz. Mater.,* 2, 143, 1985.

116. Mosey, F. E., Swanwick, J. D., and Hughes, D. A., Factors affecting the availability of heavy metals to inhibit anaerobic digestion, *Water Pollut. Control,* 70, 668, 1971.

117. Newton, J. R., Pilot-Scale Studies on Leaching from Landfills. III. Leaching of Hazardous Wastes, WLR Tech. Note Ser. No. 51, Department of the Environment, London, 1977.

118. Postgate, J. R., *The Sulphate-Reducing Bacteria,* 2nd. ed., Cambridge University Press, Cambridge, 1984.

119. Lovley, D. R. and Phillips, E. J. P., Organic matter mineralization with reduction of ferric iron in anaerobic sediments, *Appl. Environ. Microbiol.,* 51, 683, 1986.

120. Gould, J. P., Pohland, F. G., and Cross, W. H., Chemical controls on the fate of mercury and lead co-disposed with municipal solid waste, *Water Sci. Technol.,* 21, 833, 1989.

121. Robinson, H. D. and Gronow, J. R., A review of landfill leachate composition in the UK, in *Proc. Sardinia '93, 4th Int. Landfill Symp.,* S. Margherita di Pula, Cagliari, Italy, October 11–15, 1993, 1087.

122. Christensen, T. H., Albrechtsen, H. J., Kromann, A., Ludvigsen, L., and Skov, B., The degradation of chlorinated aliphatic compounds in a sanitary landfill, in *Proc. Sardinia '93, 4th Int. Landfill Symp.,* S. Margherita di Pula, Cagliari, Italy, October 11–15, 1993, 1086.

123. Craft, D. G. and Blakey, N. C., Codisposal of sewage sludge and domestic waste in landfills, Paper presented at ISWA '88 15th International Solid Wastes Conference, Copenhagen, Denmark, September 11–16, 1988.

CHAPTER 5

Landfill Leachate Treatment

Trevor J. Britz

CONTENTS

0-87371-968-9/95/$0.00+$.50

I. INTRODUCTION

One of the most important aspects that must be taken into consideration during the operation and management of landfills is their capacity to generate leachate and particular attention must be paid to the protection of surface and ground waters. In most countries landfill leachate is considered to pose a serious environmental pollution hazard and must, therefore, be disposed of in an environmentally responsible manner. Various studies and field observations indicate that even small municipal landfills may impact groundwater.[1] Since many of the newly constructed landfills are sealed at the bottom, leachate production becomes an even greater problem and must be removed for treatment by physical-chemical or biological methods.[2] The collection and treatment of leachate have a limited history but are now recognized as one of the greatest problems associated with the operation of landfill sites.[3] The field of knowledge is in a continual state of change and improvement. Today's "state of the art" system may soon be obsolete and similarly regulations vary greatly. Despite the variability, it is obvious that the trend is for most new landfills to have leachate collection capabilities. This, of course, dictates the necessity for leachate treatment and disposal alternatives. Thus, the management of landfill leachate will be a significant concern for years to come.

Because of the high-strength and labile nature of the leachate, rapid deoxygenation of the receiving watercourse can occur. Growth of sewage fungus, caused by high C:N ratios and low molecular weight carbon compounds, is often evident.[4] Ammonia in the leachate from older landfills is also a problem and will often cause deoxygenation and fish kills in the receiving waters. An aesthetic problem is the precipitation of iron compounds leading to staining.[4] Odors may be a major source of irritation and are associated with reduced sulfur compounds which appear to be an environmental nuisance rather than a toxic hazard. Survival of potential pathogenic bacteria and viruses may lead to the contamination of drinking water resources.[4,5]

Leachate is a complex organic liquid formed primarily by the percolation of precipitation water through the open landfill or through the cap of the completed site. To a lesser extent, leachate can be formed as a result of the initial moisture content of the waste.[6] The resulting leachate is a complex and highly variable mixture of soluble organic, inorganic, bacteriological constituents and suspended solids in an aqueous medium. Since all organic materials in the waste undergo partial or total microbial decomposition (mineralization), all leachates contain intermediate products together with high concentrations of toxic organics, heavy metals, and other xenobiotic materials. The exact composition is variable and site specific depending on the type and age of refuse and the amount of precipitation. It is apparent that the leachate quality and composition[7] can vary so widely that attempts to define a typical leachate must include such broad concentration ranges of the different contaminants as to be virtually meaningless for treatability studies.[8]

The most important environmental parameter affecting leachate composition is the age of the landfill. Leachate from young landfills contain high COD levels as a result of the presence of fatty acids. They also tend to be easily degradable and are characterized by high BOD:COD ratios, higher pH, and lower heavy metal concentrations.[6] As the proportion of labile organic compounds decreases with age, biological leachate treatment processes tend to become less effective.[9] The efficacy of a treatment method that works well for a young site has been observed to become progressively less effective as the site stabilizes.[10] Furthermore, the leachate properties and pollutant concentrations of one landfill may differ significantly from those of another but comparable site.[11] Because of the variability in leachate composition, leachate should be well characterized prior to the selection of a treatment process.

Proper landfill design and site management can significantly reduce the quantity and strength of leachate but will never eliminate it. Leachate management is, thus, a key consideration in landfill siting. Not only does the leachate impact the environment but also it can significantly affect the cost of designing, operating, and closing a landfill. Before selecting a new site, a workable plan for leachate disposal must be developed. Leachate disposition has the potential to become one of the costliest capital improvement and budget line items in operating a lined landfill. Because the leachate's strength can increase over time as its volume decreases, the flexibility to handle these changes should be designed into the disposition system. Furthermore, since the production of leachate may continue for many years after the landfill has ceased operations, this must be taken into consideration when sites are planned or closed. Leachate monitoring and treatment have the potential to be long-term cost and liability issues of up to 30 years after site closure.[3]

The goal of leachate treatment is to reduce the concentration of pollutants to facilitate discharge into surface waters or to pretreat constituent concentrations to acceptable levels for transfer to an off-site treatment facility. An encouraging sign in the management of leachate is the recognition by private and public site operators of the importance of assessing leachate generation at both existing and proposed sites as part of environmental planning. Since leachate characteristics vary considerably, both qualitatively and quantitatively, these will, consequently, dictate the general strategy to be adopted for successful or partial treatment of leachate in order to satisfy environmental quality directives. Most leachates cannot be treated adequately by just one method and a combination is generally recommended and, perhaps, even a final form of advanced treatment. The most important constraint is the cost of treatment due to the nature and strength of the leachate that varies daily and seasonally. Treatment plants thus need to be robust and flexible because the type of treatment will change from biological to a combination of biological and chemical as the emplaced waste ages. It is, therefore, necessary to devise treatment options which involve minimal labor and operating costs, as well as using stable treatment options.

II. TREATMENT METHODS

Leachate treatment methodologies can roughly be grouped into four main categories:

A. Leachate channeling
    1. Off-site treatment
    2. Recycling
    3. Irrigation
B. Physical and chemical treatment
    1. Chemical evaporation
    2. Chemical oxidation
    3. Chemical precipitation
    4. Activated carbon adsorption
    5. Gamma irradiation
    6. Reverse osmosis
    7. Ammonia stripping
C. Biological treatment
    1. Aerobic treatment
    2. Anaerobic treatment
D. Combined physical/chemical/biological systems

A. LEACHATE CHANNELING

1. Off-Site

If the landfill site has the advantage, by virtue of its location, of having a relatively large municipal waste water treatment plant with available capacity that is receptive to receiving leachate, it can be piped there for combined treatment with domestic sewage. In general, this is the simplest leachate management approach[12,13] and, when possible, the preferred disposal method.[14,15] This disposal option is applicable since leachate usually contains an excess of N, and sewage, an excess of P. Therefore, these nutrients, essential for microbial degradation of organic matter, would not be deficient. Pohland and Harper[16] also reported that these combined systems[15] were capable of BOD_5 and COD removal efficiencies exceeding 90%.

Lower results have also been reported[15] with the reduction of ammonia concentrations only around 80%. The addition of leachate, however, did impair COD removal and, as a result of the increased organic loads in these combined systems, plant expansions could be required.

In many cases, pretreatment standards necessitate treatment prior to sewer discharge. Usually these standards are dictated by the constituents to be removed and the economics. Consideration should be given to the oxygen transfer capacity, which must be sufficient to maintain aerobic conditions within the aeration basin with the increased organic load. Nutrients, especially phosphorus, must be present in nonlimiting concentrations to maintain the necessary operating conditions.[15] Other problems may stem from the nature of the leachates, in terms of low acidity, and high organic and inorganic concentrations, as determined by the age of the site.[10] A further consideration is the possible presence of toxic organic compounds that may not be degradable in the conventional sewage treatment plant.[10]

In other cases, sewer fees for organic waste loading have been reported to be higher than the cost of constructing and operating a pretreatment plant. Construction costs of a treatment facility must be included in the overall leachate treatment costs. Comparison of the costs of on-site and off-site treatments has indicated[4] that little economic benefit is gained by direct discharge, unless the strength of the leachate is less than 2000 mg/l COD.[17] As a general rule,[4] a leachate volume >5% of the total sewage plant input[3] and leachate COD >10 g O_2 /l is not normally acceptable[14] to a receiving works. One reason for this is because of the high ammonia concentration (5000 to 30,000 mg/l COD and 100 to 300 mg/l NH_3-N). Other problems such as toxicity to microorganisms affecting sewage digestion, diminished sludge settling, precipitation of Fe oxides, and corrosion may occur. Under these conditions, the final plant effluent may be a frothy scum and may contain high concentrations of NH_3.[18] Furthermore, the high metal content inhibits biological sludge activity and reduces or eliminates the possibility of its subsequent application as fertilizer.[3] However, in different locations[14] the amount of leachate acceptable in a municipal system differs significantly and the limiting proportion should be determined using the actual leachate.

2. Recycling

A considerable amount of research has been made to demonstrate the efficacy potential of leachate recirculation.[19,20] It appears to offer a management option that can help diminish long-term adverse environmental impacts. Recirculation back through the refuse mass substantially affects leachate treatment needs and has the potential to significantly reduce leachate contaminant concentration in a relatively short period. The debate over the desirability of recirculation has continued for years. If leachate is recirculated, biological treatment occurs within the landfill and the leachate leaving the site, therefore, requires less treatment.[12] Those opposed to the practice feel that to minimize leachate potential necessitates concerted effort to exclude liquid from the site. If the objective is to provide accelerated landfill stabilization within a more manageable and predictable time frame, and also to recover gas as a guaranteed energy source, then leachate recycling may emerge as the preferred alternative.[21]

Recirculation through the emplaced waste offers potential advantages in reducing the liquid volume by surface evaporation and reducing the leachate strength by crude anaerobic treatment. Robinson[22] found that the greatest contribution to controlling leachate volumes is the liquid storage within perched areas of the site itself.[17] Increased moisture content promotes refuse stabilization and enhances gas production.[23] Pohland[24] reported that leachate recycling enhances contact and homogeneity of reaction opportunities with a resultant decrease in environmental parameter variability and a faster attainment of stabilized conditions for biological and chemical processes. By recycling leachate, especially in arid areas, stabilization processes that normally take 30 to 50 years can be accomplished in 5 to 10 years.[25] Thus, the time required for biological stabilization[23,26] of the labile leachate compounds is generally reduced by as much as an order of magnitude.[27] Recirculation can help reduce the yearly amount by evaporation during the dry season and greater than 80% of discharged leachate has been successfully recirculated[22] with evaporation losses of between 30 and 35%.

Leachate recycling has been shown to be successful in both pilot and full-scale.[27,28] Field-scale trials have indicated[28] that recycling, with careful site selection and planning, can be used successfully as a partial option in leachate management, although a secondary route for final liquid disposal is often required. According to Henry[14] it is not clear how much decomposition of organics can be obtained. In pilot-scale studies, 97% COD reductions have been found but in larger field trials only 40% were obtained over 20 months.[22] However, it has been shown that the benefits obtained with recycling found in smaller studies can be obtained on a larger scale but longer recycle periods are required to produce low-strength leachate. Pohland[29] reported that leachate recycle effects large reductions in COD and VFA concentrations in a short time (about 300 days), while Leckie et al.[30] found that recirculation causes an initial increase in COD, followed by a decrease to negligible values in about 2 years. It has also been reported that the COD, in the presence of recirculation, drops about 70% in 3 years[31] and that the time required before stable methanogenesis commences is double in the absence of recirculation. Matta-Alvarez and Martinez-Viturtia[32] developed a model[3] to stimulate site dynamics with leachate recycle and showed that more than 90% of biodegradable matter is converted in 25 to 60 days at temperatures of 34 to 38°C and that the active micro-biological life of the site can be reduced to <2 years.

One distinct advantage of the increased moisture content as result of the recirculation is that methane production is optimized and the higher gas volumes are recoverable and usable. However, Barlaz et al.[33] found that there were no significant differences in the total anaerobic population or in the subpopulations of cellulolytic, hemicellulolytic, acetogenic, or meth-anogenic bacteria in refuse, incubated with or without leachate recycling. Differences in soluble constituent concentrations and methane production patterns between leachate recycle and nonrecycle were attributed to the mixing associated with leachate recycling.

It is also necessary to control the pH of the leachate and a lime treatment as part of the recirculation process for partial neutralization and metal re-moval[31] is recommended. The wide deviations in pH can reduce the anaer-obic microbial population, thus lengthening the biological stabilization process.[3] pH control at a neutral value also facilitates the more rapid es-tablishment of an active methanogenic population and subsequent biogas production. Improvements in pH control and initial sludge seeding[27] may further enhance treatment efficiency and reduce the biological stabilization period.

The hydrology of the site should also be carefully considered,[22] espe-cially how this is affected by use of cover material.[23] Measures taken to ensure the correct distribution of recycled leachate are important to prevent the formation of channels and lateral movement into surrounding surface or ground water. It has also been shown[22] that surface ponding can be alleviated by furrowing. Unfortunately, the conventional methods of con-struction and operation of landfills with daily cover do not necessarily fa-cilitate optimum hydraulic characteristics for leachate recycle within the refuse mass.[25,34] Pohland[27] also found that control of the leachate as the major transport phase provides new opportunities for evaluation, design, and control for co-disposal sites.

In certain areas, recirculation may not be possible if the prior water balance indicates an accumulation of liquid in the site. When liquid does not accumulate, this is one of the least expensive options available.[3] Current evidence suggests lower costs of leachate recycle in contained sites as com-pared to direct biological options.[27] Unfortunately, recirculation alone cannot provide a final solution for leachate management. Although the leachate organic fraction can be reduced by recycling, other constituents are not significantly removed (NH_4^+, Ce^+, and metals in particular)[4] and thus recirculation cannot be seen as a complete solution. Since the final effluent generally does not meet recommended disposal standards,[10,22] a further treatment step must be considered when direct discharge to the en-vironment is contemplated.

3. Irrigation

The concept of leachate land spraying revolves around the usage of adjacent lands[10] as a natural medium for biological and physical-chemical processes. This method is recommended by certain workers[10,17] but others find it unacceptable.[3] According to Lema,[3] land spraying is not a valid treatment and is generally seen as a noxious practice. The main reason for this[3] is that high concentrations of toxic elements present in leachate can make land unfit for agricultural use and courts the risk of polluting surface and groundwater.

According to Saint-Fort,[10] irrigation offers an economical and viable alternative option, provided that suitable areas are available. However, before use, the influence of each component of the leachate on the soil must be assessed. Irrigation of adjacent land or even a completed landfill has proved to be an effective process[4] which is widely practiced in the U.K. where it has been found to be a low-cost option. Hydraulic loadings of 56 m³/ha/d have been reported to be acceptable and, at a loading of 45 m³/ha/d, BOD reductions of 95% have been achieved.[4] From the literature it is clear that irrigation is more suitable for low-strength leachate and cannot be used during the colder months[4] when overspraying could lead to waterlogged conditions.

In the presence of vegetation cover, further volume reductions by plant uptake and transpiration should be effected.[35] During summer periods, Robinson[17] also found that it was possible to evaporate and transpire substantial quantities of leachate by spraying onto various grasslands. Unfortunately, high-strength leachate, even at low flow rates, caused damage to the vegetation. In contrast, Gordon et al.[36] concluded that the irrigation of red maple seedlings with leachate at rates consistent with evapotranspirational demand of the plants resulted in few undesirable short-term effects, but, at application rates that give rise to reduced conditions, a 100% mortality can occur. Other workers[37] reported that irrigation with leachate increased plant biomass production significantly, but this depended on the quality and quantity of the leachate as well as the type of plant. It was also reported[38] that short-rotation plantations can successfully be established at landfills still in operation in order to control leachate volumes by evapotranspiration. It has also been shown that, if leachate application rates are not regulated in accordance with evapotranspiration demand and precipitation regimes, excess soil water and poor soil aeration will persist and inevitably lead to reduced conditions and vegetation die-back.[36] Furthermore, if leachate properties indicate that a precipitate will develop, then pretreatment will be necessary.[36] This may be a problem during winter periods when high leachate volumes have to be treated in the presence of low evapotranspiration rates and low microbial activity.[34] Gordon[36] also reported that the irrigation of leachate has a negative effect on the total microbial biomass present in soil. A decrease in microbial biomass was also noted with increasing soil depth in areas that were spray irrigated, in comparison with unsprayed controls. The application of leachate to forest soil may cause a shift in speciation of the microbial community possibly due to changes in the microbial decomposition and nutrient cycling processes.

In addition to possible soil structure damage,[34] the presence of high concentrations of heavy metals may result in phytotoxicity to vegetation and plant die-back since effective adsorption of toxic metals such as lead, cadmium, and nickel does not obviate plant uptake.[34,39] Thus, a pretreatment might have to be undertaken prior to discharge. Even though it is known that soils do have the capacity[14] to reduce organics and adsorb metals, there is resistance to using irrigation as a treatment option. Due consideration, before using land spraying, should also be given to factors such as soil type, presence of cations, moisture content, presence of organic matter, depth of ground water, nutrient additions, and microbial inocula.[34] The main disadvantages are the large areas needed and the scarcity of information on the long-term effects of applying leachate to soil. A further drawback is the possibility of odor production in the immediate vicinity of the irrigated areas.[34]

B. PHYSICAL AND CHEMICAL TREATMENT

1. Chemical Evaporation

Evaporation studies[40,41] have shown that satisfactory results can be obtained with leachate. It was found that two-stage distillation, with a high pH followed by an acid step, resulted in low organic and metal concentrations in the final distillate. This, however, represents a very expensive treatment option.

2. Chemical Oxidation

The purpose of chemical oxidation has been defined by Saint-Forte[10] as a method to render leachate contaminants insoluble, gasify them, or stabilize them as relatively innocuous substances. Even though limited literature exists[42] regarding the chemical oxidation of leachate, several processes have been examined. These include wet oxidation,[43,44] ozonation, peroxide treatment, and chemical reduction.[45]

Ho et al.[46] studied the effects of chlorine (Cl_2), ozone (O_3), calcium hypochlorite [$Ca(ClO)_2$] and potassium permanganate ($KMnO_4$). The COD reduction in leachate for all four chemicals, even at high doses, was less than 48%. However, excellent iron and color removal were obtained, although with the treatment of calcium hypochlorite it was found that the hardness of the final solution increased. With chlorine,[8,42,47,48] low COD removals of between 20 and 30% have generally been found. Similarly, Loizidou et al.[42] found that with hydrogen peroxide (30% w/w) a reduction of about 35% was achieved, with only about 20% of the ammonia being oxidized. Thus, it was concluded that hydrogen peroxide is best suited for treating dilute streams and must be handled carefully.[44] One advantage is that the removal of hydrogen sulfide may be effected by the addition of 1.5 parts H_2O_2 to 1 part sulfide at a pH >8.5 and contact time of 30 to 60 min.[49] Harrington and Maris.[4] considered hydrogen peroxide primarily as a remedial option and suggested its usage for both bad odor control and treatment.

It has been reported that,[42,50] for leachate pretreatments, oxidants such as hydrogen peroxide and potassium permanganate successfully reduce nickel concentrations, but the leachate is likely to foam. Excess foaming, especially on full-scale treatment, would be unacceptable. Kang et al.[50] also reported that, in addition to foaming, potassium permanganate treatment generated 10 to 50% sludge by volume.

Due to solubility and mass transfer limitation and generation costs, ozone is only applicable to waste streams containing less than 1% oxidizable materials.[44] Both ozone and hydrogen peroxide are preferable to chlorination for oxidation of leachates since no chlorinated by-products are produced.[44] Chemical oxidants are often the most expensive to use and will only be cost effective with low BOD:COD ratio leachate and low concentrations of organics.[6] These methods are energy intensive and require special precautions and may well find application in treating biologically recalcitrant leachates.[44] The use of halogenated oxidants has been reported to generate highly toxic organohalogen compounds.[3,51]

3. Chemical Precipitation

Several methods for the chemical precipitation of compounds in leachates have been tested and include lime—$Ca(OH)_2$; alum—$Al_2(SO_4)$; ferric chloride—$FeCl_3$; sodium sulfide—Na_2S; and ferrous sulfate—$FeSO_4$.[10,52] During precipitation, the major goals are the removal of heavy metals and a portion of the organic matter.[53] High heavy metal concentrations can inhibit biological activity or exceed downstream requirements[6] and As, Cu, Zn, Cd, Pb, Ni, Ag, Hg, Cr, and Fe are usually targeted for control. Sedimentation and filtration processes[54] are also used to remove suspended solids, stabilized sludge, or precipitate from leachate. The removal of the solids will reduce the organic loading on a biological system. Sufficient time must be given to allow solids to settle and, when combined with flocculant addition (alum or polyelectrolytes), will cause particles to coagulate and settle faster.[6] In terms of COD removal, precipitation is not efficient and is reported[3,4] to be less than 30% and, thus, not considered an adequate treatment option on its own.[10] Good improvements in color, suspended solids, NH_4^+, and heavy cation elimination are obtained.

Extensive literature[18,55] is available on chemical precipitation as a pretreatment technique.[44,52,56–59] The increase in pH as a result of lime addition[60] leads to the coagulation[53] and formation of insoluble metal hydroxides and calcium carbonate. The resulting flocs aid in the settling of colloidal material. Lime treatment alone, using up to 1000 mg/l $Ca(OH)_2$, has been reported to be negative in terms of the leachate COD characteristics,[61] but when combined with potassium permanganate a substantial COD reduction of 59% was obtained. The effectiveness of metal precipitation when using lime depends[53] on the efficiency of separating the precipitate from the waste stream. Precipitation when combined with sand filtration has been shown to result in metal reduction of 60%.[62] Boiler ash[63] has also been reported as a pretreatment method with excellent removal efficiency of metals and about 33% COD removal.

The use of aluminum precipitation ($AlSO_4$) has been studied on laboratory scale and in full scale.[64] Satisfactory purifications were obtained, but the proportion of leachate in combined treatment should not exceed 2%. Alum combined with a polyelectrolyte gave a 31% COD removal, while a combination with lime resulted in a 39% COD removal.[65]

Precipitation of heavy metals can be a complex chemical process with chelant and electrolyte interactions causing wide variations[6] from theoretical predictions. The advantages[4] of applying physical and chemical methods include flexibility and fast start-up periods and reaction rates, thus reducing the plant size compared to biological treatments.[34] They are usually insensitive to temperature and many of the methods lend themselves to automation.[4] The disadvantages are that physical-chemical options[4] are often the most expensive, operating costs can be high, and sludge is generated that must be removed. Sludges from the processes are potentially hazardous[6,10] and should be disposed according to regulations. For complete treatment, physical-chemical processes are inadequate as most leachates cannot be treated effectively by only conventional physical-chemical methods.

4. Activated Carbon Adsorption

Activated carbon adsorption is a well-developed treatment option,[10] with reports on good removals of most organics.[62] However, removal is ineffective for organics such as acetone and methanol.[62] According to Enzminger et al.,[44] carbon adsorption is the most extensively used physical-chemical means of removal of organic constituents from leachate. The best results were obtained when combined with biological methods,[52,66] even on full scale. The main disadvantage is the need for frequent regeneration of the carbon columns or an equivalently high consumption of carbon powder.[3] Spent carbon is regenerated by hot air stripping of the adsorbed organics and subsequent incineration. Adsorption can also be combined with filtration as a pretreatment to prevent plugging of carbon adsorption beds and to further reduce the solids levels. Back-washable sand filters are usually employed.[6]

The adsorptive capacity depends on the preparation method, storage conditions, pore size, surface area, and solution pH.[44] Handling and energy costs are high,[6] making this option cost effective only for the removal of residual organics but only when the total dissolved solids in the solution are less than 200 mg/l.[18] This method is generally only used for treating leachate from old sites or as a tertiary treatment of the final effluent from biological units where a final COD reduction of 85% can be achieved. However, it is a simple and effective method and requires low capital inputs.[6]

5. Gamma Irradiation

The decomposition of refractory substances in leachate, using radiation-induced oxidation under aerobic conditions, has been examined.[34,67,68] It was found that the total organic carbon decreased with eventual conversion to carbon dioxide. An outstanding characteristic of the radiation treatment is the ability to decompose most of the organics without producing sludge.[68] For high organic concentrations, equivalent high radiation dosages are required, thus making this a very expensive treatment option. A combination of gamma radiation (79 kGy at pH 3) and ferric chloride coagulation under anaerobic conditions has been reported to be less expensive and very effective,[68] with about 80% reduction in organics.

6. Reverse Osmosis

Several researchers have investigated the possibility of using reverse osmosis for the treatment of leachates[52,69–71] and high organic removals, in excess of 90%, were achieved. Krug and McDougall[70] also successfully used a microfiltration-reverse osmosis process with more than 95% organic and inorganic removals. Chian and DeWalle[70,72] observed that reverse osmosis treatment of leachate was the most effective physical-chemical method for COD removal. According to Kettern,[73] most treatment plants in Germany are now being planned with a reverse osmosis unit.

With reverse osmosis, organics in leachate were reported to be reduced by 97% and total solids by 97%.[74] Metals as high as 860 µg/l were reduced to below detection limits and the permeate was of extremely high quality. An optimum pressure for reverse osmosis operation was found to be 250 psi on pilot-scale in the field in a facility capable of treating 10,000 gallons per day. The concentrate was recycled back to the site.[74] Similarly, it has been reported that reverse osmosis leachate treatment[74] resulted in a permeate stream of extremely high quality and a concentrated waste stream.

As Cd, Cr, Cu, Pb, Ni, and Zn concentrations in permeate scans were below detection limits. It was found that better results were obtained when the leachate was treated soon after collection and before any biological activity took place in the holding tanks. If not, hydrogen sulfide odors developed in the permeate.[74] The buildup of gas in older leachate as a result of biological activity also presented problems at the start of the reverse osmosis process. This problem could also be eliminated by treating the leachate soon after collection. In another pilot and full-scale study,[75] 92% of the influent stream was converted to permeate and 95 to 99% removal of organics, nitrogen, phosphate, and metals resulted. However, chemical cleaning of the membranes had to be included as a regular part of the operation. Chian and DeWalle[52] recommended that an appropriate pretreatment of leachate prior to reverse osmosis processing would help to prevent severe membrane fouling.[70]

Results reported by Weber and Holz[76] revealed distinct advantages for biological treatment prior to reverse osmosis. The application of reverse osmosis after biological intervention was also found to remove 80% of the organics in leachate.[77] Excellent results were also reported by Van Der Schroeff[78] for an UASB reactor, (operating capacity 2 m³/h), followed by a reverse osmosis unit, operating on a landfill site.

The reverse osmosis treatment method suffers from membrane fouling and prolongation of membrane life necessitates elimination of suspended solids and colloidal material. The type of membrane, pH, pressure, and pretreatment are important aspects in determining the effectiveness of the reverse osmosis treatment process.[72] However, treatment costs[73] relate only to physical separation of the leachate constituents. On average, 20% of the raw leachate is produced as a concentrate[73] which has to be passed on for further disposal. By recycling to the site the problem is merely shifted elsewhere.

7. Ammonia Stripping

Leachates from refuse that has been emplaced for longer periods are generally weaker but contain relatively high concentrations of ammonia-nitrogen,[4] which is toxic and must be regulated in the final discharge water. Raising the pH of the leachate to above pH 9 with lime, in order to form NH_3 gas, and bubbling air through the system, can accomplish ammonia removal for atmospheric discharge of up to 100%.[3,18,55,79] Air stripping as a chemical method can thus be used to control excessive concentrations of ammonia.

Full-scale use of the air-stripping method has been evaluated,[80] while results using tower and lagoon stripping have been reported.[4,49] Keenan[53] combined air stripping with a lagoon system (950 m³) and reported ammonia removals of up to 50%. Smith and Arab[79] reported an almost 100%

removal on laboratory scale with leachate containing 570 mg/l NH_3-N. Prasad et al.[81] reported a 90% removal of the total nitrogen after 8 h aeration at a pH of 10, a temperature of 20°C, and an average air flow rate of 150 cm^3/s/l. A final solution containing between 2.2 and 5.2 mg/l ammonia was produced. Prasad[81] also studied the influence of environmental factors on ammonia removal and described a practical model for predicting ammonia removal by air stripping. Optimum air stripping[10,79] is strongly influenced by temperature, wind speed, aeration rate, lagoon configuration, pH control, surface area, and initial and required final concentration in the leachate.

Air stripping is generally considered a cost-effective and unsophisticated system, but stripping towers can be expensive to build and operational problems include the formation of an adherent scale in the tower.[79] In contrast, lagoons are seen as a low-cost alternative, which are particularly desirable for on-site treatment. A typical air-stripping plant consists of a lagoon, an aeration system, and a pH control and feeding unit.[10] The high pH is also advantageous in removing heavy metals by precipitation and then air stripping the ammonia from the leachate,[14] thus minimizing ammonia and metal inhibitory effects.

C. BIOLOGICAL TREATMENT

Biological degradation is one of the most promising options for the removal of organic material from leachate. However, sludge formed, especially during the aerobic biodegradation processes, may become a serious disposal problem. This can be aggravated as a result of the ability of sludge to adsorb specific organic compounds and toxic heavy metals. However, biological systems have the advantage of microbial transformations of complex organics and possible adsorption of heavy metals by suitable microbes.[44] Biological processes are still fairly unsophisticated but, according to Enzminger et al.,[44] great potential for combining various types of biological schemes for selective component removal exists.

1. Aerobic Biological Systems

a. Aerobic Reactors

Aerobic biological treatment methods depend on microorganisms grown in an oxygen-rich environment to oxidize organics to carbon dioxide, water, and cellular material. Considerable information[14] on laboratory and field-scale aerobic treatments of leachate is available,[2,4,8,72,82,83] but little has been published on full-scale aerobic systems[3,54] although several plants are in operation.[73,84]

Aerobic treatment has been shown to be reliable and cost effective in producing a high-quality effluent. Start-up usually requires an acclimation period to facilitate the development of a competitive microbial community. Ammonia-nitrogen in leachates from young landfills can successfully be removed in order to prevent subsequent disposal problems.[14,22,85] Problems normally associated with aerobic processes include foaming and poor solid-liquid separation.[3] Phosphorus additions (BOD:N:P = 100:3:1 or 100:3.2:1.1) are also necessary[3,14] and sludge volumes generated (1 kg SS/kg BOD removed) are about double the values from municipal wastes. These may also contain high levels of heavy metals and will require disposal as hazardous wastes.[6]

Activated Sludge Plants — Lema et al.[3] summarized several published results of successful research using sludge plants with COD removals varying from 30 to 99% under different operational and loading conditions. Keenan et al.[53] used an activated sludge plant consisting of aeration tanks and secondary clarifiers, and also found the method to be successful, both in series or parallel operational modes, with BOD_5, COD, and ammonia removals of 99, 95, and 90%, respectively. Stegmann[2,54] suggested that as a guideline, in order to obtain complete degradation and to produce an effluent with <25 mg/l BOD_5 in an activated sludge plant, the volumetric organic load must be less than 0.15 kg $BOD_5/m^3/d$.

With activated sludge plants, problems generally encountered are foam production, precipitation of iron and carbonates, excessive sludge production (0.6 kg dry sludge/kg of BOD_5 removed), and a decrease in efficiency during winter periods.[54] According to Maris et al.,[86] conventional activated sludge processes are less appreciated as they demand greater operator skill and a degree of skill not normally found at landfill sites.

Aerobic Filters — Conventional trickling filters,[87] with aerobic microbes growing on rock or gravel, are limited in depth to about 2 m, as deeper filters enhance anaerobic growth with subsequent odor problems.[54] In contrast, filters with synthetic media are fully aerobic up to about 8 m.

Knox[85] used a conventional field-scale trickling filter to treat old leachate with low BOD_5 and high NH_3 concentrations. The HRT varied between 15 h and 4.5 days, with NH_3-N removal varying between 36 and 309 g N /unit surface area (m^2). The filter had fewer operational problems than activated sludge systems and was simple to operate for the treatment of leachates with low $BOD_5:NH_3$ ratios.[54] However, an inherent problem is that the trickling filters can be blocked by precipitated ferric hydroxide[34,49] and carbonates, with concomitant reduction of microbial activity.[88] Maris et al.[86] reported that biological filters are not appropriate for the treatment of high-strength leachates as filter blinding by organic deposition on the filter medium is generally found. However, Pedersen and Jansen,[89] using a pilot-scale submerged aerobic biological filter to treat a high ammonia leachate, found a maximum nitrification capacity of 0.57 kg NH_4-$N/m^3/d$ at 20°C, 66% suspended solids, 81% organic matter, and 88% COD removal.

Continuous-Flow Aerobic Reactors — Aerobic continuous-flow reactors have been shown[39] to stabilize leachate effectively and good COD (92%) and BOD (97%) removals can be achieved at 5 to 10°C with phosphate addition at solid retention times of 20 days. At lower temperatures, however, the process was adversely effected.

Rotating Biological Contactors (RBC) — Smith and Moore[90] compared an RBC with a conventional activated sludge process for the detoxification of a hazardous leachate waste stream. The removal efficiencies ranged from 50 to 95% for both organic and toxic constituents. The RBC achieved higher organic and heavy metal removals and showed a greater resilience to organic and toxic shock loads than the activated sludge system. In subsequent studies they showed[91] that organic loads of 1.59 to 1.54 $kg/m^3/d$ can be achieved with 95 to 99% BOD_5 removal. They also reported chlordane removal of 75 to 96% in the same system.

Coulter[92] successfully used the RBC on pilot-scale to treat leachate-containing phenols, iron, and cyanide with 98, 97, and 92% removal efficiencies, respectively, and a BOD_5 removal of up to 98%. The treatability of low BOD (26 mg/l) and high ammonia (154 mg/l) leachate from an old site was also examined by Spengel and Dzombak.[93] At steady state a removal of 80% BOD_5, 38% COD, and 98% ammonia was found. However, low substrate loadings were used (1.2 to 7.3 g NH_3-$N/m^3/d$ and 2.8 to 18.5 g $COD/m^3/d$).[93] Thus, a significant amount of nondegradable COD (also responsible for the color of the leachate) remained after treatment and further steps would be needed if effluent limits were imposed for color or COD loads. Similar findings have been reported.[86,94,95]

The RBC process appears to offer several advantages over the activated sludge process[93,96,97] for use in leachate treatment especially those from old sites.[4] The primary advantage is the relative ease of operation and maintenance.[98] Furthermore, pumping, aeration, and wasting/recycle of solids are not required, leading to less operator attention. Operation for nitrogen removal is also relatively simple and routine maintenance involves only inspection and lubrication.

Sequencing Batch Reactors (SBR) — The SBR utilizes the same tank to aerate, settle, and recycle solids[6] and is seen as a good option with low flow applications. Irvine et al.[99] found the SBR to be a cost-effective option with TOC removals of 85 to 95%. On pilot-scale,[100] SBRs have also proved efficient,[101] with BOD removals exceeding 99%, nitrogen removals varying

from 75 to 95%, and excess phosphorus removed, by simultaneous precipitation, to less than 1.5 ppm.[101] The studies of Ying et al.[102] led to the
construction and successful operation of a full-scale plant with more than
90% TOC removal, as well as the removal of all phenolic compounds (800
mg/l) present in the leachate.

b. Lagoons/Ponds/Reed Beds

In areas where land is available, aerated lagoons constitute one of the
least expensive methods of biological treatment.[54] As a result of their sim-
plicity and absence of a sludge recycle facility, lagoons are a favored
method for effective leachate treatment.[86,103] Aerated lagoons are easy to
install and require relatively unskilled attention. Floating aerators may be
used and evacuation on the site plus lining is a simple method of lagoon
construction.[86] Although lagoons are easy to operate, they are the most
complex of all biologically centered degradation systems.[104] In these systems, both aerobic and anaerobic metabolisms occur in addition to photo-
synthesis and sedimentation.

Operators of sites in warmer climates[6] may find the use of lagoons a
suitable and economical leachate treatment option. However, the potential
does exist for surface and ground water pollution, bad odors, and insects
that may become a nuisance.

Chian and DeWalle[72] found that, when treating a strong leachate (COD
57,900 mg/l) in an aerated lagoon, organic removal of between 93 and 97%
could be obtained without pretreatment, at retention times of 7 to 86 days
and organic loadings of 0.244 to 1.65 kg/m^3/d. Heavy metal reductions were
also reported. The COD:P ratios were found to be critical in the presence
of phosphorus limitation. Maris et al.[86] reported that even simple aeration
methods removed readily degradable constituents at temperatures as low as
5°C. Unless phosphorus is added, aerobic treatment is slow.[86] The removal
of fatty acids was accompanied by 85% COD reduction and up to 98%
reduction in BOD, with about 10% reduction in ammonium-nitrogen.[86] Thirumurthi[54] also reported excellent BOD removal of 98% but found lower
COD (79%) and TOC (73%) removals as a result of the presence of recalcitrant organics. Only 36% of the NH_4-N was converted but after combining
with an aeration tank a removal >99% was obtained.[54] Smith[105] used a
combined cyclic aeration and settlement process and obtained significant
improvement in ammonium removal.

Field studies[17,106,107] have shown that aerated lagoons can be operated
successfully during severe winter months at retention times of longer than
10 days,[4] while treating up to 150 m^3/d leachate to a high standard, with
minimal staff attention. In Germany, many sites include aerated lagoons

with foam control and phosphorus addition.[108] Complete oxidation and nitrification were reported by Stegmann,[109] but at hydraulic times of longer than 100 days.

Extended aerated lagoons are now an established technology in the U.K. and have been shown[108] to be a robust, stable, and reliable means of treating leachates. Ammonia-nitrogen removals in excess of 1000 mg/l have been reported.[108] Other aerobic systems lack the substantial dilution and buffering[109] which an aerated system provides against short-term variations in influent quality and flow at ambient temperatures. The lagoon system lends itself to simple automation and requires minimal attention from on-site staff.[109]

Full-scale studies in the U.S., Canada, U.K., and Europe have shown that aerated lagoons can effectively treat leachate. In the U.K. alone, ten such automated systems were in operation in 1993.[110] Many more such aerated biological systems are under design and construction, both in the U.K. and elsewhere, on the basis of their proven performance records and robustness.[110] The recommended design parameters vary considerably and depend on the quality of the raw leachate, the required degree of degradation, and the operational temperature.[54] Waste treatments by lagoons or ponds are classified on their mode of biodegradation (aerobic or anaerobic), the presence or absence of aeration equipment, and other design features.[104] The classifications and kinetics of lagoons and ponds have been summarized by Thirumurthi.[104]

Reed bed pond systems have also found widespread application.[107] A design manual and operating guidelines were produced in 1990.[107,111] Reed beds are designed to treat leachate by passing it through rhizomes of the common reed in a shallow bed of soil or gravel. The reeds introduce oxygen and as the leachate percolates through aerobic microbial communities establish among the roots and degrade contaminants.[108] Nitrogen and phosphorus are probably removed directly by the reeds.[108] Reed beds, however, are poor in removing ammonia and, with high concentrations of ammonia being toxic, this may be a limiting factor. The precipitations of large quantities of iron, manganese, and calcium within the reed beds also affect rhizome growth and, in time, reduce the permeability of the bed. According to Robinson et al.,[107] field studies in the U.K. have shown that reed beds have enormous potential and, in combination with aerobic systems, provide high effluent quality at reasonable cost.[112] This process could later, at older sites, provide a low-cost, long-term treatment option[108] when used in isolation.

c. Nitrification/Denitrification

Leachates from older landfills are generally high in ammonia and low in biodegradable carbon and effect receiving water through ammonia toxicity, by providing nutrients for eutrophication and by exerting an oxygen demand through nitrification. Biological nitrification-denitrification is considered one of the most suitable methods for removing ammonia from leachate.[14,113] However, due to the low carbon content, an external carbon source must be added[114] to facilitate effective denitrification. According to Manoharan et al.,[113] many carbon wastes are suitable but they must be carefully chosen.[115] When using glucose in a single-sludge system, with leachate containing 170 to 230 mg/l ammonia, Manoharan[113] reported inconsistent denitrification (10 to 100%), but with methanol as carbon source denitrification was found to be complete. The innovative use of a jam factory effluent to control the imbalance in the carbon-ammonia-nitrogen ratio was reported by Robinson[116] and resulted in an excellent final effluent quality.

Denitrification of leachate with biogas (methane content 50%) as sole carbon source is also feasible.[117] Similarly, maximum denitrification rates of 550 mg NO_3-N/l/d were obtained in a fluidized reactor, with 30 to 40% methane in the gas phase.[117] Biogas can thus be seen as a long-standing, reliable, readily available, and inexpensive carbon source to achieve complete denitrification. Together with essential carbon supplementation, oxygen and CO_2 concentrations play significant roles in the process optimization.[117]

Denitrification efficiencies of over 90% have been reported.[118] Field studies, using activated sludge and trickling filter pilot plants,[85] showed that ammonia removal was successful at between 13 and 16°C, with a removal rate of over 300 mg $N/m^3/d$. Imai et al.[119] evaluated the long-term performance (700 days) of a fluidized process in removing refractory organics and nitrogen from leachate and found 60 and 70% removal, respectively.

In bench-scale studies the usefulness of both aerobic and anaerobic rotating biological contactors (RBCs) in treating low BOD (26 mg/l), low COD (358 mg/l COD), and high ammonia (154 mg/l NH_3-N) leachate from an older site was evaluated. Attached microbes in the aerobic contactor removed 80% BOD, 38% COD, and 98% of the ammonia. With the anaerobic system it was demonstrated that the nitrate produced by nitrification could successfully be removed. A kinetic model was also developed for these systems.[10] Knox[85] also reported that the RBC was successful for treatment of ammonia-containing leachates.

Nitrification-denitrification treatment systems exposed to metals capable of precipitating phosphorus should be monitored to ensure sufficient nutrient availability.[120] Manoharan et al.[120] recommend a minimum soluble phosphorus concentration of 0.5 mg/l. Nitrification of ammonia in leachates is likely to be of increasing importance, as more and more landfill sites are designed to reach a methanogenic state.[85] Nitrification should present no problem, provided that an active nitrifying population is present and the environmental conditions for these autotrophic organisms are suitable.[118] However, high concentrations of metals could inhibit the process.

2. Anaerobic Biological Treatment

The anaerobic treatment process depends on a microbial association growing under anaerobic conditions and, subsequently, converting the organic compounds to methane, carbon dioxide, and other metabolites. It has often been stated[121] that anaerobic methods for leachate treatment are too slow, unreliable, and sensitive to pH variations as well as to the presence of heavy metals. This is contradictory to the literature in which Boyle and Ham[8] stated that anaerobic digestion is one of the most promising treatment options. According to a summary by Blakey and Maris,[121] high-strength leachates are a prime substrate for methane fermentation. It has only been with the introduction of high-rate digesters that the anaerobic treatment of leachate has been considered an alternative to aerobic treatment.

Anaerobic systems are generally seen[18] as more economical for biological stabilization of leachate as they do not have the high energy requirements associated with aeration in aerobic systems. Anaerobic digestion also yields methane which can be utilized as a heat or power source. Furthermore, less sludge is generated thereby reducing problems associated with its disposal. Nutrient requirements (N and P) are much lower than for aerobic systems,[54] pathogenic organisms are usually destroyed, and the final sludge has a high soil conditioning value if the concentrations of heavy metals are low. The possibility of treating high COD leachate without previous dilution, as required by aerobic systems, reduces space requirements and thus the associated costs.[3] Bad odors are generally absent if the system is operated efficiently. Recent literature, however, indicates that many toxic organic compounds can be biotransformed under anaerobic conditions.[54]

The disadvantages associated with anaerobic systems are the high capital cost, long start-up periods, strict control of operating conditions, greater sensitivity to variable loads and organic shocks, as well as toxic compounds.[122] The operational temperature must be maintained at about 33 to 37°C for efficient kinetics while it is important to keep the pH at a value around 7, as a result of the sensitivity of the methanogenic population to low values.[54] Heavy metal removal can be good but is not as efficient as in aerobic processes. Since ammonia-nitrogen is not removed in an anaerobic system, it is consequently discharged with the digester effluent, thus creating an oxygen demand in the receiving water. Therefore, there is also the need for complementary treatment to achieve acceptable discharge standards.

As part of the anaerobic process, hydrogen sulfide is formed and this provides an efficient precipitant for most metals. These precipitates accumulate as inert solids within the sludge blanket[123] or on the filter media. Similarly, calcium is precipitated as carbonate or phosphate salts and also accumulates in the system and, as with the metal precipitates, must be removed from the digester.[10,121,123]

Problems associated with phosphate as a nutrient in anaerobic systems have been studied extensively by Lema et al.[3] It was found that, at low organic loadings, a leachate with less than 1 mg/l PO_4 could be treated efficiently. Similar results were reported by Henry et al.[9] and Mendez et al.[7] However, Lema et al.[3] did report that at higher loadings phosphate addition was necessary. Henze and Harremoes[124] proposed a theoretical range of COD:N:P varying from 2000:7:1 to 400:7:1 to support efficient anaerobic digestion. According to Thirumurthi,[125] this is a conservative range. Thirumurthi[125] while using a filter design, found that by adding PO_4, an increase in COD removal resulted. He reported that at an organic loading of 8 kg $COD/m^3/d$ and HRT of 2.6 days, the maximum sustainable COD:P ratio in the feed leachate was 30,300:1. The corresponding minimum total phosphate in the influent was 0.7 mgP/l. A 91% COD removal could be achieved when the COD to P ratios were between 21,600 and 30,300:1. Decreases in COD removal were found with ratios above 30,300:1. Based on these results, Thirumurthi[125] recommended that, for economic reasons, the minimum PO_4 concentration required for successful treatment must be determined before final digestion systems are designed.

a. Anaerobic Reactors

Conventional Digesters — The conventional process utilizes mechanical or gas mixing within the reactor to provide contact between the microorganisms and the substrate. Sludge recovery is by settlement and it is sometimes recycled back to the digester. Boyle and Ham[8] in 1974, while using anaerobic conventional design, reported a COD removal of 90 to 96% at loadings of 0.43 to 2.2 kg $COD/m^3/d$ at HRTs of 5 to 20 days and temperatures of 23 to 30°C. Other workers using the same design also reported good operational efficiencies.[126-128] Cossu[129] described the use of a simple anaerobic lagoon which effected COD removals of 80 to 90% after an extended retention time of 40 to 50 days at 25°C. The process was reported to be very sensitive to temperature variations, even with extended retention times.

Anaerobic Filters — The anaerobic filter consists of a flooded reactor, packed with support media with the flow being either upward or downward. Loading rates depend on the amount of active biomass that can be retained in the reactor. The biomass concentration is directly proportional to the biofilm thickness and the total surface area for biofilm development.[130] The filter offers several advantages which include low sludge production, relatively low capital and operating costs, a simple design, and easy operation.[130]

The efficiency of leachate treatment by anaerobic methods was demonstrated in 1977 by Chian and DeWalle,[72] using a completely mixed anaerobic filter with feedback. They reported efficient COD removals over a wide range of organic loading rates. Since the initial work by Young and McCarty,[28,131] anaerobic filters have received considerable attention as a viable option to treat high-strength leachates.[7,52,130–137]

Mendez et al.[7] evaluated in laboratory scale the use of upflow anaerobic filters with recycle for treating leachates from young and old sites at HRTs of 1.85 to 10.67 d. At a maximum loading rate of 1.92 kg COD/m^3/d and HRT of 1.85, a COD removal of 36% and methane yield of 0.313 m^3/kg COD$_{removed}$ was obtained. Henry et al.[138] also found no appreciable differences in terms of COD and volatile solids removal, gas production, and the methane content between upflow and downflow filters or between plastic biorings and stone media filters. Henry et al.[9,138] reported reductions in COD of more than 90%, using the anaerobic filter for the treatment of different age leachates at loading rates of 1 to 2 kg COD/m^3/d at room temperature, with HRTs of 24 to 96 h, without phosphorus addition. In contrast, Muthukrishnan and Atwater[130] found a higher phosphorus requirement (COD:P = 115:1 to 330:1) for anaerobic treatment of leachate. Kinetic constants have also been determined to study volatile fatty acids removal in an anaerobic filter[132] treating leachate. Carter et al.[134] also successfully operated a full-scale anaerobic filter treatment system treating leachate.

A modification of the anaerobic filter with no recycle, to include a sludge bed at 35°C,[139] resulted in a soluble COD removal of more than 90% at organic loadings of up to 13 kg COD/m^3/d and a HRT of up to 216 d. The removals of sulfate and soluble Fe obtained were 90 and 96%, respectively. About 68% of the sludge solids generated were accumulated in the system. Similar studies by Kennedy and Guiot[140] at loadings of 33 kg COD/m^3/d gave COD removals of 95%. They also reported that this design could withstand severe organic shock loads and recover in a reasonable time. A pilot-scale sludge-bed filter reactor[141] was found to be capable of removing more than 85% COD at organic loadings of up to 10 kg/m^3/d. However, extended operation of the pilot plant resulted in the accumulation of significant amounts of inorganic precipitates, causing severe operational problems. Sludge activity decreased due to mass transfer limitations caused by metal precipitation in and around the sludge granules. Pretreatment[141] was recommended for the removal of the metals, which form insoluble precipitates under anaerobic conditions.

Fixed Film Reactors — Kennedy et al.[142] successfully treated leachate with a downflow stationary film reactor and achieved a 94% COD removal at an organic loading of 14.7 kg COD/m^3/d and a HRT of 1.5 d.

UASB Types —Berueta and Castrillon[143] used an upflow anaerobic sludge blanket (UASB) to treat a leachate containing high concentrations of ammonia, at an organic loading of 6.8 kg COD/m³/d and HRT of 2.4 days and obtained 88% COD removal and a methane yield of 0.340m³/kg COD$_{removed}$. It was, however, also necessary to acidify continuously and this may represent an economic drawback in large-scale operations. Anaerobic pretreatment in a UASB resulted[144] in COD removals of more than 90% at loadings of 9.4 kg COD/m³/d and more than 99% removal of Ca, Fe, Cd, Cu, Pb, and Zn. Nutrient addition was unnecessary at the organic loadings used. Similar results, with UASB reactors, have also been reported by several other workers.[28,123,145–147]

Kennedy et al.[142] used an anaerobic upflow blanket filter at a loading rate of 14.7 kg COD/m³/d and maintained a 97% COD removal at a HRT of 1.5 d. An inorganic heavy metal precipitate was present in the sludge bed and filter media.[125,148] Raw leachate was treated in this system[142] at loading rates of up to 44 kg COD/m³/d and HRTs down to 10 h with COD removals in excess of 88%. The maximum specific removal rate was 1.2 kg COD/kg VSS/d. These hybrid designs combine the advantages of the upflow sludge blanket systems with the fixed film/filter systems, while minimizing the disadvantages.[142]

Britz et al.[149] treated a high-strength leachate (COD = 18 kg/l) using a UASB-fixed-bed digester at a HRT of 1 d and 35°C and loading rate of 18.6 kg COD/m³/d. A COD removal of 82% was obtained with a methane yield of 0.281 m³/kg COD$_{removed}$. They reported that system overloading was characterized by the accumulation of large quantities of volatile acids, especially propionic acid. Myburg and Britz[150] used the same design but after a longer conditioning period and showed that higher organic loading rates (26.05 kg COD/m³/d) can successfully be used with a 95% COD removal and a methane yield of 0.215 m³/kg COD$_{removed}$.

Pilot-scale UASB studies in the U.K.[121] have indicated that up to 85% BOD removal can be achieved with an average organic loading of about 12 kg COD/m³/d at an average HRT of 1.75 d and a biogas methane content of 70 to 80% (v/v). Thirumurthi[54] summarized the operational data of a full-scale leachate treatment plant,[137] which was constructed as a combination of upflow sludge blankets/filter units. The anaerobic component of the treatment plant was designed to remove 90% of the raw leachate COD.

D. COMBINED PHYSICAL/CHEMICAL/BIOLOGICAL SYSTEMS

For the purification of industrial wastewaters, combined anaerobic-aerobic processes have become more popular.[151,152] A substantial part of the organic load can be removed anaerobically, thereby producing methane. The remaining pollution load is then removed aerobically, using aeration energy. This combined technology is one of the most economical carbon removing solutions. It has been demonstrated that combined aerobic-anaerobic biological systems can also be effectively used to treat leachates.[152,153] Venkataramani et al.[153] recommended that an anaerobic pretreatment, followed by an aerobic polishing step, provides the optimal approach. Combined anaerobic-aerobic systems have also been shown to be successful[154] for the complete degradation of most chlorinated aliphatic, propene, and ethane compounds.

Biological treatments have also been combined successfully with physical-chemical methods.[27,71,155,156] Keenan et al.[53] concluded that no one treatment method separately achieves a high enough removal efficiency. They advocated that a combined strategy should be employed where physical-chemical methods are used to remove metals and hydrolyze part of the organic fraction, and biological methods are employed to stabilize the degradable organic matter. Dienemann et al.[152] recommended reverse osmosis as an attractive post-treatment after a combined aerobic-anaerobic process. A fixed-film anaerobic treatment, after a precipitation step, followed by an aerated lagoon,[157] resulted in >99% removal of COD, BOD, TOC, Fe, and Zn. The biogas generated ranged from 0.40 to 0.57 m^3/kg $COD_{removed}$. Similarly, Ho et al.[46] reported that it was possible to remove between 95 and 98% COD in a combined aerobic-anaerobic system, followed by lime precipitation. Other biological-chemical combinations[54] also resulted in BOD, COD, and ammonia removals of 95 to 99, 95, and 89 to 99%, respectively.[53,158,159] The BOD:COD ratio of the final effluent was 0.16:1.

In certain cases, physical-chemical treatments[6] have been used as final polishing steps after biological treatment but the choice of strategy is particularly important.[34] In view of the complexity of leachate treatment problems, the most effective option must be tailored to suit the site and in many cases may be combined effectively with other treatment processes.

One serious problem encountered during leachate treatment is ammonium-nitrogen removal. Physical-chemical methods can also be used as a pretreatment step to remove ammonium-nitrogen before biological treatment is applied.[49] However, denitrification in the aerobic stage is usually incomplete due to the low carbon content. In order to achieve total denitrification,[151] external carbon sources must often be added, resulting in a substantial cost. To eliminate the extra cost, Abeling and Seyfried[151] used a new process of nitrification-denitrification via nitrite with a 40% lower carbon consumption. Control of the pH and ammonium content were essential in order to maintain process stability.

III. CONCLUSIONS

Management of landfill leachates will be an ever-increasing concern in the future and treatment strategies will have to be based on national and local regulations. Before selecting any leachate treatment method, a complete process evaluation, together with an economical analysis, should be undertaken. This should include the leachate composition, concentrations, volume, and treatment susceptibility as well as the environmental impacts of the strategy to be adopted. As Kennedy et al.[142] stated, all options are expensive but an economical analysis may indicate that slightly higher maintenance costs may be cheaper than increased operating costs. What is appropriate for one site may be unsuitable for another.

The most useful processes are those that can be operated with a minimum of supervision and are inexpensive to construct or even mobile enough to be moved from site to site. The changing nature of leachate (quantity and quality) must also be included in the design and operational procedures. From the literature it appears as if biological methods are the most cost effective for the removal of organics, with aerobic methods being easier to control, although anaerobic methods have lower energy requirements and lower sludge production rates. Physical-chemical methods have been shown to be effective in removing heavy metals and colloidal matter. However, they do not destroy partitioned constituents with the result that further processing or disposal is necessary. Since no single process for treatment of leachate is capable of complying with the minimum effluent discharge requirements, it may be necessary to choose a combined process especially designed to treat a specific leachate.

REFERENCES

1. Bagchi, A., Natural attenuation mechanisms of landfill leachate and effects of various factors on the mechanisms, *Waste Manage. Res.*, 5, 453, 1987.
2. Stegmann, R., New aspects on enhancing biological processes in sanitary landfill, *Waste Manage. Res.*, 1, 201, 1983.
3. Lema, J. M., Mendez, R., and Blazquez, R., Characteristics of landfill leachates and alternatives for their treatment: a review, *Water Air Soil Pollut.*, 40, 223, 1988.
4. Harrington, D. W. and Maris, P. J., The treatment of leachate: a UK perspective, *Water Pollut. Control*, 85, 45, 1986.
5. Gray, M., De Leon, R., Tepper, B. E., and Sobsey, M. D., Survival of hepatitis A virus (Hav), poliovirus 1 and F-specific coliphages in disposable diapers and landfill leachates, *Water Sci. Technol.*, 27, 429, 1993.
6. Harris, J. M. and Gaspar, J. A., Management of leachate from sanitary landfills, *Aiche Symp. Ser. NY*, No. 265, 84, 171, 1988.
7. Mendez, R., Lema, J. M., Blazquez, R., Pan, M., and Forjan, C., Characterization digestibility and anaerobic treatment of leachates from old and young landfills, *Water Sci. Technol.*, 21, 145, 1989.
8. Boyle, W. C. and Ham, R. K., Biological treatability of landfill leachate, *J. Water Pollut. Control Fed.*, 46, 860, 1974.

9. Henry, J. G., Prasad, D., and Young, H., Removal of organics from leachates by anaerobic filter, *Water Res.,* 21, 1395, 1987.

10. Saint-Fort, R., Fate of municipal refuse deposited in sanitary landfills and
leachate treatability, *J. Environ. Sci. Health,* A27, 369, 1992.

11. Knox, K., The Relationship between Leachate and Gas, Paper presented at
Energy and Environment Conference, Bournemouth, England, 1990.

12. Schaper, L. T., Trends in landfill planning and design, *Public Works,* April,
64, 1991.

13. Niininen, M., Kalliokoski, P., and Eskelinen, T., Co-treatment of landfill
leachate and domestic sewage in activated sludge plant: a case study in
Finland, in Int. Conf. *Environ. Pollut. Proc.,* Switzerland, 1991, 307.

14. Henry, J. G., New developments in landfill leachate treatment, *Water
Pollut. Res. J. Can.,* 20, 1, 1985.

15. Kelly, H. G., Pilot testing for combined treatment of leachate from a domestic waste landfill site, *J. Water Pollut. Control Fed.,* 59, 254, 1987.

16. Pohland, F. G. and Harper, S. R., Critical Review and Summary of Leachate
and Gas Production from Landfills, Report to the EPA, WERL, Coop.
Agreement CR809997, Cincinnati, OH, 1985.

17. Robinson, H. D., Design and operation of leachate control measures at
Compton Bassett landfill site, Wiltshire, U.K., *Waste Manage. Res.,* 5, 107,
1987.

18. Lisk, D. J., Environmental effects of landfills, *Sci. Total Environ.,* 100, 415,
1991.

19. Baetz, B. and Onysko, K. A., Storage volume sizing for landfill leachate-
recirculation systems, *J. Environ. Eng.,* 119, 378, 1993.

20. Al-Yousfi, A. B., Pohland, F. G., and Vasuki, N. C., Design of landfill
leachate recirculation systems based on flow characteristics, in *47th Purdue
Industrial Waste Conference Proceedings,* Lewis Publishers, Chelsea, MI,
1992, 491.

21. Pohland, F. G., Dertien, J. T., and S. B. Ghosh., Leachate and gas quality
changes during landfill stabilization of municipal refuse, in Proc. 3rd Int.
Symp. Anaerobic Digestion, Boston, 1983, 185.

22. Robinson, H. D. and Maris, P. J., The treatment of leachates from domestic
waste in landfill sites, *J. Water Pollut. Control Fed.,* 57, 30, 1985.

23. Barber, C. and Maris, P. J., Recirculation of leachate as a landfill management option: benefits and operational problems, *Q. J. Eng. Geol.,* 17, 19,
1984.

24. Pohland, F. G. and Gould, J. P., Testing methodologies for landfill codisposal of municipal and industrial wastes, in *Hazardous and Industrial Solid
Waste Testing and Disposal,* Vol. 6, Lorenzen, D., Conway, R. A., Jackson,
L. P., Hamza, A., Perket, C. L., and Lacy, W. J. Eds., ASTM Special Technical Publication, Baltimore, MD, 1986, 45.

25. Lee, G. F. and Jones, R. A., Landfills and ground-water quality, *Ground
Water,* 29, 482, 1991.

26. Senior, E., Landfill: the ultimate anaerobic bioreactor, Proc. 1st South African Anaerobic Digestion Symposium, Bloemfontein, South Africa, 1986,
230.

27. Pohland, F. G., Harper, S. R., Chang, K.-C., Dertien, J. T., and Chian, E. S. K., Leachate generation and control at landfill disposal sites, *Water Pollut. Res. J. Can.,* 20, 10, 1985.

28. Young, C. P., Blakey, N. C., and Maris, P. J., Leachate management, in *Harwell Landfill Practical Symp. Proc.,* Energy Technology Support Unit, Harwell Laboratories, Harwell, Oxon, U.K., May, 1987.

29. Pohland, F. G., Landfill Stabilization with Leachate Recycle, Annual Progress Report. EPA No. EP-00658, Washington, DC, 1972.

30. Leckie, J. D., Pacey, J. G., and Halvadakis, C. P., Landfill management with moisture control, *J. Eng. Div. Am. Soc. Civ. Eng.,* EE2, 337, 1979.

31. Boari, G., Mancini, I. M., and Spinosa, L., Landfill leachate: Operating modalities for their optimal treatment, in CEE, WP1, paper presented at Workshop on Thermophilic Digestion and Landfilling of Sludge, Nancy, France, November 1987.

32. Matta-Alvarez, J. and Martinez-Viturtia, P., Laboratory simulation of municipal solid waste fermentation with leachate recycle, *J. Chem. Technol. Biotechnol.,* 36, 547, 1986.

33. Barlaz, M. A., Ham, R. K., and Schaefer, D. M., Microbial, chemical and methane production characteristics of anaerobically decomposed refuse without leachate recycling, *Waste Manage. Res.,* 10, 257, 1992.

34. Senior, E. and Shibani, S. B., Landfill leachate, in *Microbiology of Landfill Sites,* Senior, E., Ed., CRC Press, Boca Raton, FL, 1990, 81.

35. Ettala, M., Evapotranspiration from a *Salix aquatica* plantation at sanitary landfill, *Aqua Fennica,* 18, 3, 1988.

36. Gordon, A. M., McBride, R. A., and Fisken, A. J., Effect of landfill leachate irrigation on Red Maple (*Acer. rubrum L.*) and Sugar Maple (*Acer. Saccharum Marsh.*) seedling growth and on foliar nutrient concentrations, *Environ. Pollut.,* 56, 327, 1989.

37. Ettala, M., Influence of irrigation with leachate on biomass production and evapotranspiration on a sanitary landfill, *Aqua Fennica,* 17, 69, 1987.

38. Ettala, M., Short-rotation tree plantations at sanitary landfills, *Waste Manage. Res.,* 6, 291, 1988.

39. Robinson, H. D. and Maris, P. J., The treatment of leachates from domestic wastes in landfills. I. Aerobic biological treatment of a medium-strength leachate, *Water Res.,* 17, 1537, 1983.

40. Leonhard, K., Eindampfung von deponiesickerwasser, *Wasser Abwasser,* 131, 467, 1990.

41. Andretta, F., Fractional distillation for treating toxic leachates, *Asian Water Sewage,* 29, 46, 1990.

42. Loizidou, M., Papadopoulos, A., and Kapetanios, E. G., Application of chemical oxidation for the treatment of refractory substances in leachates, *J. Environ. Sci. Health,* A28, 385, 1993.

43. Dietrich, M. J., Randall, T. L., and Ganrey, P. J., Wet air oxidation of hazardous organics in wastewater, *Environ. Progr.,* 4, 171, 1985.

44. Enzminger, J. D., Robertson, D., Ahlert, R. C., and Kosson, D. S., Treatment of landfill leachates, *J. Haz. Mater.,* 14, 83, 1987.

45. Tucker, S. P. and Carsen, G. A., Deactivation of hazardous chemical waste, *Environ. Sci. Technol.,* 19, 215, 1985.

46. Ho, S., Boyle, W. C., and Ham, R. K., Chemical treatment of leachate from sanitary landfills, *J. Water Pollut. Control Fed.,* 46, 1776, 1974.

47. Cook, E. N. and Forre, E. G., Aerobic biostabilization of sanitary landfill leachate, *J. Water Pollut. Control Fed.,* 46, 380, 1974.

48. Karr, R. P., Treatment of Leachate from Sanitary Landfills. Special Research Problems, Report of the School of Civil Engineering, Georgia Institute of Technology, Atlanta, 1972.

49. Robinson, H. D. and Maris, P. J., Leachate from Domestic Waste: Generation, Composition and Treatment—A Review. Technical Report, U.K. Water Research Centre, Stevenage, U.K., 1979, 108.

50. Kang, S. J., Englert, C. J., Astfalk, T. J., and Young, M. A., Treatment of leachate from a hazardous waste landfill, in *44th Purdue Industrial Waste Conference Proceedings,* Lewis Publishers, Chelsea, MI, 1990, 573.

51. Gould, J. P. and Ramsey, R. E., Formation of volatile halorganic compounds in the chlorination of municipal landfill leachates, in *Water Chlorination Environment Impact and Health Effects,* Vol. 4, Jolly, R. A., Ed., Ann Arbor Science, Ann Arbor, MI, 1983, 525.

52. Chian, E. S. K. and DeWalle, F. B., Sanitary landfill leachates and their treatment, *J. Environ. Eng. Div.,* 102, 441, 1976.

53. Keenan, J. D., Steiner, R. L., and Fungaroli, A. A., Landfill leachate treatment, *J. Water Pollut. Control Fed.,* 56, 27, 1984.

54. Thirumurthi, D., Biodegradation of sanitary landfill leachate, in *Biological Degradation of Wastes,* Martin, A. M., Ed., Elsevier, London, 1991, 208.

55. Venkataramani, E. S., Ahlert, R. C., and Carbo, P. Biological treatment of landfill leachates, *Crit. Rev. Environ. Control,* 14, 333, 1984.

56. Slater, C. S., Uchrin, C. G., and Ahlert, R. C., Physicochemical pretreatment of landfill leachates using coagulation, *J. Environ. Sci. Health,* A18, 125, 1983.

57. Davis, R. A., The use of magnesium hydroxide to reduce metal hydroxides sludge filtration and disposal cost, in Proc. Haz. Proc. Conference, Baltimore, MD, 1985, 405.

58. Kim, B. M., Treatment of metal containing wastewaters with calcium sulfide, *Am. Inst. Civ. Eng. Symp. Ser.,* 77, 39, 1981.

59. Ying, W.-C., Duffy, J. J., and Tucker, M. E., Removal of humic acid and toxic organic compounds by iron precipitation, *Environ. Progr.,* 7, 262, 1988.

60. Sharma, V. K. and Sapienza, F. C., Design development of a landfill leachate pretreatment plant, in *Proc. Natl. Waste Processing Conference,* Amercian Society of Mechanical Engineers, New York, 1990, 313.

61. Kinman, R. N. and Nutini, D. L., Physical/chemical treatment of sanitary landfill leachate, in *46th Purdue Industrial Waste Conference Proceedings,* Lewis Publishers, Chelsea, MI, 1991, 1.

62. Liu, B. Y. M., Li, A., Henry, M. P., and Kruckel, B., Evaluation of a full-scale physico-chemical treatment plant for removal of bdat constituents from hazardous waste landfill leachate, in *45th Purdue Industrial Waste Conference Proceedings,* Lewis Publishers, Chelsea, MI, 1991, 599.

63. Gray, M. N., Rock, C. A., and Pepin, R. G., Pretreating landfill leachate with biomass boiler ash, *J. Environ. Eng.,* 114, 465, 1988.

64. Ettala, M., Chemical precipitation of leachate from sanitary landfills at municipal sewage treatment plants, *Aqua Fennica,* 24 19, 135, 1989.

65. Loizidou, M., Vithoulkas, N., and Kapetanios, E., Physical chemical treatment of leachate from landfill, *J. Environ. Sci. Health,* A27, 1059, 1992.

66. Ying, W.-C., Bonk, R. R., and Sojka, S. A., Treatment of a landfill leachate in powdered activated carbon enhanced sequencing batch bioreactors, *Environ. Progr.,* 6, 1, 1987.

67. Yamazaki, M., Sawai, T., and Sawai, T., Radiation treatment of landfill leachate, *Radiat. Phys. Chem.,* 18, 76, 1981.

68. Sawai, T., Sawai, T., and Yamazaki, M., Effect of gamma irradiation on the removal of fulvic acid in landfill leachate by coagulation, *Radioisotopes,* 38, 51, 1989.

69. Slater, C. S., Ahlert, R. C., and Uchrin, C. G., Treatment of landfill leachate by reverse osmosis, *Environ. Progr.,* 2, 251, 1983.

70. Krug, T. A. and McDougall, S., Preliminary assessment of a microfiltration/ reverse osmosis process for the treatment of landfill leachate, in *43rd Purdue Industrial Waste Conference Proceedings,* Lewis Publishers, Chelsea, MI, 1988, 72.

71. Kettern, J. T., Landfill leachate purification: state-of-the-art, in *Int. Conf. Environ. Pollut. Proc.,* Switzerland, 1991, 314.

72. Chian, E. S. K. and DeWalle, F. B., Evaluation of Leachate Treatment, Vol. 2, Biological and Physical-Chemical Processes, EPA Report 600/277–186b, Cincinnati, OH, 1977.

73. Kettern, J. T., Biological or chemico-physical landfill leachate purification—an alternative or a useful combination, *Water Sci. Technol.,* 26, 137, 1992.

74. Kinman, R. N. and Nutini, D. L., Reverse osmosis treatment of landfill leachate, in *45th Purdue Industrial Waste Conference Proceedings,* Lewis Publishers, Chelsea, MI, 1991, 617.

75. Bilstad, T. and Madland, M. V., Leachate minimization by reverse osmosis, *Water Sci. Technol.,* 25, 117, 1992.

76. Weber, B. and Holz, F., Significance of the biological pretreatment of sanitary landfill leachate on the reverse osmosis process, in *2nd Int. Landfill Symposium,* Porte Conte, Sardinia, Italy, 1989.

77. Venkataramani, E. S. and Ahlert, R. C., Acclimated mixed microbial responses to organic species in industrial landfill leachate, *J. Haz. Mater.,* 10, 1, 1985.

78. Van Der Schroeff, J. A. M., Post-Treatment of Anaerobic Treatment: a Grown-up Technology, Report, Amsterdam. IPG, Schiedam, The Netherlands, 359, 1986.

79. Smith, P. G. and Arab, F. K., The role of air bubbles in the desorption of ammonia from landfill leachates in high pH aerated lagoon, *Water Air Soil Pollut.,* 38, 333, 1988.

80. Steiner, R., Keenan, J., and Fungaroli, A., Demonstration of a Leachate Treatment Plant, U.S. National Technical Information Service, Springfield, VA, PB/269/502, 1977.

81. Prasad, D., Henry, J. G., and Elefsiniotis, P., Nitrogen removal from anaerobically treated leachate, *Water Pollut. Res. J. Can.,* 20, 138, 1985.

82. Knox, K., Treatability studies on leachate from a co-disposal landfill, *Environ. Pollut.*, 5, 157, 1983.
83. Ehrig, H. J., Treatment of sanitary landfill leachate: biological treatment, *Waste Manage. Res.*, 2, 131, 1984.
84. Robinson, H. D. and Gronow, J. R., A review of landfill leachate composition in the UK, in Proc. 4th Int. Landfill Symp., Sardinia, Italy, 1993, 821.
85. Knox, K., Leachate treatment with nitrification of ammonia, *Water Res.*, 19, 895, 1985.
86. Maris, P. J., Harrington, D. W., Biol, A. I., and Chismon, G. L., Leachate treatment with particular reference to aerated lagoons, *Water Pollut. Control*, 83, 521, 1984.
87. Bakdwubm W. C. and Ellis, N. H., Packaged system treats landfill leachate, *Pollut. Eng.*, 19, 44, 1987.
88. Stegmann, R. and Ehrig, H.-J., Operation and design of biological leachate treatment plants, *Water Technol.*, 12, 919, 1980.
89. Pedersen, B. M. and Jansen, J. L. C., Treatment of leachate-polluted groundwater in an aerobic biological filter, *Eur. Water Pollut. Control*, 2, 40, 1992.
90. Smith, J. W. and Moore, L. W., Biological Detoxification of Hazardous Waste, Report of the Tennessee Water Resources Research Center, Knoxville, 2, 1984.
91. Smith, J. W., Moore, L. W., and Sabatini, D. A., Biological treatment of simulated landfill leachate containing chlordane, in 18th Mid-Atlantic Ind. Waste Conf. Proc., Blacksburg, VA, 1986, 157.
92. Coulter, R. G., Pilot-plant study of landfill leachate treatment by biological contactors, *Pollution Equipment News*, USPS/349/750, 1985, 18.
93. Spengel, D. B. and Dzombak, D. A., Biokinetic modelling and scale-up considerations for rotating biological contactors, *Water Environ. Res.*, 64, 223, 1992.
94. Peddie, C. and Atwater, J. W., RBC treatment of a municipal landfill leachate: a pilot scale evaluation, *Water Pollut. Res. J. Can.*, 20, 115, 1985.
95. Hartmann, K.-H. and Hoffmann, E., Leachate treatment: design recommendations for small but extremely fluctuating, highly polluted quantities of water, *Water Sci. Technol.*, 22, 307, 1990.
96. Lugowski, A., Haycock, D., Poisson, R., and Beszedits, S., Biological treatment of landfill leachate, in *44th Purdue Industrial Waste Conference Proceedings*, Lewis Publishers, Chelsea, MI, 1990, 565.
97. Opatken, E. J., Howard, H. D., and Bond, J. J., Stringfellow leachate treatment with RBC, *Environ. Progr.*, 7, 41, 1988.
98. Lambert, M. E., Columbus RBC research project: utility manager's viewpoint, in *Proc. 2nd Int. Conf. on Fixed Film Biological Processes*, Arlington, VA, 1984, 849.
99. Irvine, R. L., Sojka, S. A., and Colaruotolo, J. F., Biological treatment of leachate from an industrial landfill: specialized bacteria, in *Ann. Haz. Mater. Manage. Conf. Proc.*, Wheaton, IL, 1984, 159.
100. Ying, W.-C., Bonk, R. R., Lloyd, V. J., and Sojka, S. A., Biological treatment of a landfill leachate in sequencing batch reactors, *Environ. Progr.*, 5, 41, 1986.

101. Morling, S. and Johansson, B., Sequencing batch reactor (SBR) tests on concentrated leachates at low temperature, Varberg, Sweden, *Vatten,* 45, 223, 1989.

102. Ying, W.-C., Wnukowski, J., Wilde, D., and McLeod, D., Successful leachate treatment in SBR-adsorption system, in *47th Purdue Industrial Waste Conference Proceedings,* Lewis Publishers, Chelsea, MI, 1992, 501.

103. Frigon, J.-C., Bisaillon, J.-G., Paquette, G., and Beaudet, R., Caracterisation et traitement du lixiviat d'un lieu d'enfouissement sanitaire, *Sci. Techniq. De L'eau,* 25, 469, 1992.

104. Thirumurthi, D., Biodegradation in waste stabilization ponds (facultative lagoons), in *Biological Degradation of Wastes,* Martin, A. M., Ed., Elsevier, New York, 1991, 231.

105. Smith, P. G., Removal of ammonia from landfill leachates and other wastewaters, *Public Health Eng.,* 12, 159, 1984.

106. Robinson, H. D. and Grantham, G., The treatment of landfill leachates in on-site aerated lagoon plants: experience in Britain and Ireland, *Water Res.,* 22, 733, 1988.

107. Robinson, H. D., Barr, M. J., Formby, B. W., and Moag, A., The treatment of landfill leachates using reed bed systems, IWEM Annual Training Day, Shrewsbury, U.K., October 1, 1992.

108. Robinson, H. D., Barr, M. J., and Last, S. D., Leachate collection, treatment and disposal, *J. Inst. Water Environ. Manage.,* 6, 321, 1992.

109. Stegmann, R., Design and construction of leachate treatment plants in W. Germany, in *Symp. Landfill Leachate,* Harwell, U.K., 1982, 1.

110. Robinson, H. D., Leachate technology, *Wastes Manage.,* August, 30, 1993.

111. European Water Pollution Control Association, European Design and Operations Guidelines for Reed-Bed Treatment Systems. Report to EC/EWPCA Treatment Group, Cooper, P. F., Ed., Brussels, August, 1990, 1.

112. Robinson, H. D., The treatment of landfill leachates using reed bed systems, in Proc. 4th Int. Landfill Symp. Sardinia, Italy, 1993, 907.

113. Manoharan, R., Liptak, S., Parkinson, P., and Mavinic, D., Denitrification of a high ammonia leachate using an external carbon source, *Environ. Technol. Lett.,* 10, 701, 1989.

114. Dedhar, S. and Mavinic, D. S., Ammonia removal from a landfill leachate by nitrification and denitrification, *Water Pollut. Res. J. Can.,* 20, 126, 1985.

115. Carley, B. N. and Mavinic, D. S., The effects of external carbon loading on nitrification and denitrification of a high-ammonia landfill leachate, *Res. J. Water Pollut. Control Fed.,* 63, 51, 1991.

116. Robinson, H. D., Full scale treatment of high ammonia landfill leachate, 1988 Joint CSCE—ASCE, *Natl. Conf. Environ. Eng. Proc.,* Vancouver, 1988, 1.

117. Werner, M. and Kayser, R., Denitrification with biogas as external carbon source, *Water Sci. Technol.,* 23, 701, 1991.

118. Jasper, S. E., Mavinic, D. S., and Atwater, J. W., Influent constraints on treatment and biological nitrification of municipal landfill leachate, *Water Pollut. Res. J. Can.,* 20, 57, 1985.

119. Imai, A., Iwami, N., Matsushige, K., Inamori, Y., and Sudo, R., Removal of refractory organics and nitrogen from landfill leachate by the microorganism-attached activated carbon fluidized bed process, *Water Res.*, 27, 143, 1993.

120. Manoharan, R., Harper, S. C., Mavinic, D. S., Randall, C. W., Wang, G., and Marickovich, D. C., Inferred metal toxicity during the biotreatment of high ammonia landfill leachate, *Water Environ. Res.*, 64, 858, 1992.

121. Blakey, N. C. and Maris, P. J., Methane Recovery from the Anaerobic Digestion of Landfill Leachate. Water Research Centre, Contractor Report for the Department of Energy, U.K.—ETSU B 1223, 1990.

122. Britz, T. J., Van Der Merwe, M., and Riedel, K.-H. J., Influence of phenol additions on the efficiency of an anaerobic hybrid digester treating landfill leachate, *Biotechnol. Lett.*, 14, 323, 1992.

123. Maris, P. J., Harrington, D. W., and Mosey, F. E., Treatment of landfill leachate; management options, *Water Pollut. Res. J. Can.*, 20, 25, 1985.

124. Henze, M. and Harremoes, P., Anaerobic treatment of wastewater in fixed-film reactors—a literature review, *Water Sci. Technol.*, 15, 1, 1983.

125. Thirumurthi, D., Minimum concentration of phosphate for anaerobic fixed film treatment of landfill leachate, *Water Pollut. Res. J. Can.*, 25, 59, 1990.

126. Cameron, R. D. and Koch, F. A., Trace metals and anaerobic digestion of leachate, *J. Water Pollut. Control Fed.*, 52, 282, 1980.

127. Bull, P. S., Evans, J. V., Wechsler, R. M., and Cleland, K. J., Biological technology of the treatment of leachate from sanitary landfills, *Water Res.*, 17, 1473, 1983.

128. Lin, C.-Y., Anaerobic digestion of landfill leachate, *Water SA*, 17, 301, 1991.

129. Cossu, R., Laboratory investigation of leachate treatment by anaerobic lagooning, *Ing. Ambientale*, 13, 226, 1984.

130. Muthukrishnan, K. and Atwater, J. W., Effect of phosphorus addition on treatment efficiency of a lab scale anaerobic filter treating landfill leachate, *Water Pollut. Res. J. Can.*, 20, 103, 1985.

131. Wu, Y. C., Hao, O. J., Ou, K. C., and Scholze, R. J., Treatment of leachate from a solid waste landfill site using a two-stage anaerobic filter, *Biotechnol. Bioeng.*, 31, 257, 1988.

132. Gourdon, R., Comel, C., Martel-Naquin, P., and Veron, J., Validation of a protocol for kinetic study of VFA removal under simulated conditions of landfill leachate treatment on anaerobic filter, *Water Res.*, 26, 927, 1992.

133. Carter, J. L., Curran, G. M., Schafer, P. E., Janeshek, R. T., and Woelfel, G. C., A new type of anaerobic design for energy recovery and treatment of leachate wastes, in *39th Purdue Industrial Waste Conference Proceedings*, Lewis Publishers, Chelsea, MI, 1984, 369.

134. Carter, J. I., Schafer, P. E., Janesher, R. T., and Woelfel, G. C., Effects of alkalinity and hardness on anaerobic digestion of landfill leachate, in *40th Purdue Industrial Waste Conference Proceedings*, Lewis Publishers, Chelsea, MI, 1985, 621.

135. Thirumurthi, D. and Groskopf, G. R., Phosphate requirement for anaerobic fixed film treatment of landfill leachate, *Can. J. Civil Eng.*, 15, 334, 1988.

136. Skladany, G. J., Onsite biological treatment of an industrial landfill leachate: microbiological and engineering considerations, *Haz. Waste Haz. Mater.*, 6, 212, 1989.

137. Wright, P. J. and Austin, T. P., Detail designs of a high rate anaerobic treatment facility for a landfill leachate, in *8th Canadian Waste Manage. Conf. Proc.*, University of Hull Publisher, Hull, Quebec, Canada, 1986, 367.

138. Henry, J. G., Prasad, D., Scarcello, J., and Hilgerdenaar, M., Treatment of landfill leachate by an anaerobic filter. II. Pilot studies, *Water Pollut. Res. J. Can.*, 18, 45, 1983.

139. Chang, J.-E., Treatment of landfill leachate with an upflow anaerobic reactor combining a sludge bed and a filter, *Water Sci. Technol.*, 21, 133, 1989.

140. Kennedy, K. J. and Guiot, S. R., Anaerobic upflow bed-filter—development and application, *Water Sci. Technol.*, 18, 71, 1986.

141. Keenan, P. J., Iza, J., and Switzenbaum, M. S., Inorganic solids development in a pilot-scale anaerobic reactor treating municipal solid waste landfill leachate, *Water Environ. Res.*, 65, 181, 1993.

142. Kennedy, K. J., Hamoda, M. F., and Guiot, S. G., Anaerobic treatment of leachate using fixed film and sludge bed systems, *J. Water Pollut. Control Fed.*, 60, 1675, 1988.

143. Berueta, J. and Castrillon, L., Anaerobic treatment of leachates in UASB reactors, *J. Chem. Technol. Biotechnol.*, 54, 33, 1992.

144. Rumpf, M. I. and Ferguson, J. F., Anaerobic pretreatment of a landfill leachate for metals and organics removal, *Natl. Conf. Environ. Eng.*, 552, 1990.

145. Eng, S. C., Fernandes, X. S., and Paskins, A. R., Biochemical effects of administering shock loads of sucrose to a laboratory-scale anaerobic (UASB) effluent treatment plant, *Water Res.*, 20, 789, 1986.

146. Blakey, N. C. and Maris, P. J., On-site leachate management—anaerobic processes, ISWA—Int. Sanitary Landfill Symp. Proc., Cagliari, Sardinia, Italy, 1987.

147. Bekker, P. and Kaspers, H., Anaerobic Treatment of Leachate from Controlled Tips of Municipal Solid Waste, Paper presented at the 5th European Sewage and Refuse Symposium, EAS, Rome, June, 1981, 24.

148. Mennerich, S., Anaerobic treatment of leachate using fixed film and sludge bed systems, *Res. J. Water Pollut. Control Fed.*, 61, 1739, 1989.

149. Britz, T. J., Venter, C. A., and Tracey, R. P., Anaerobic treatment of municipal landfill leachate using an anaerobic hybrid digester, *Biol. Wastes*, 32, 181, 1990.

150. Myburg, C. and Britz, T. J., Influence of higher organic loading rates on the efficiency of an anaerobic hybrid digester while treating landfill leachate, *Water SA*, 19, 319, 1993.

151. Abeling, U. and Seyfried, C. F., Anaerobic-aerobic treatment of high-strength ammonium wastewater—nitrogen removal via nitrite, *Water Sci. Technol.*, 26, 1007, 1992.

152. Dienemann, E. A., Kosson, D. S., and Ahlert, R. C., Evaluation of serial anaerobic/aerobic packed bed bioreactors for treatment of a superfund leachate, *J. Haz. Mater.*, 23, 21, 1990.

153. Venkataramani, E. S., Ahlert, R. C., and Corbo, P., Aerobic and anaerobic treatment of high-strength hazardous liquid wastes, *J. Haz. Mater.*, 17, 169, 1988.
154. Long, J. L., Stensel, H. D., Ferguson, J. F., Strand, S. E., and Ongerth, J. E., Anaerobic and aerobic treatment of chlorinated aliphatic compounds, *J. Environ. Eng. Am. Soc. Civ. Eng.*, 119, 1, 1993.
155. Ahlert, R. C. and Kosson, D. S., Treatment of hazardous landfill leachates and contaminated groundwater, *J. Haz. Mater.*, 23, 331, 1990.
156. Albers, H., and Kayser, R., Landfill leachate and mine drainage, in *42nd Purdue Industrial Waste Conference Proceedings,* Lewis Publishers, Chelsea, MI, 1987, 893.
157. Thirumurthi, D., Austin, T. P., Ramalingaiah, and Khakhria, S., Anaerobic/ aerobic treatment of municipal landfill leachate, *Water Pollut. J. Can.*, 21, 8, 1986.
158. Keenan, J. D., Steiner, R. L., and Fungaroli, A. A., Chemical-physical leachate treatment, *J. Env. Eng. Div.*, 109, 1371, 1983.
159. Aynagiotou, C., Papadopoulos, A., and Loizidou, M., Leachate treatment by chemical and biological oxidation, *J. Environ. Sci. Health*, A28, 21, 1993.

CHAPTER 6

Landfill-Covering Soils

Chris A. du Plessis and Jeff C. Hughes

CONTENTS

0-87371-968-9/95/$0.00+$.50

I. INTRODUCTION

In landfills, soil has traditionally been used as a material for both the daily and final covering of refuse. This was, and is, seen to be an efficient way of containing odors and, more particularly, improving site appearance. This chapter will show that intermediate and final landfill-covering soils are also of great importance because of their particular interactions with the pollutant chemicals in leachate and gas. These interactions, and an understanding of them, are vital when considering landfills as either potential environmental pollutant sources or bioreactors which facilitate attenuation of key substances. Although some of these interactions also occur among pollutants, microorganisms, and the refuse mass, this discussion will be limited to soil. Interactions among soil, microorganisms, and liquid phase pollutants (leachate) have received the most attention in the literature, although specific interactions with landfill gases have also been reported.[1,2]

A better understanding of these interactions will be of great practical importance for optimal use and management of landfills and their subsequent reclamation. The interactions generally discussed in this respect mainly consider the major physical interactions such as hydraulic conductivity and the mobility of various pollutants in soil according to their water solubility and charge.[3] The interactions which will be discussed here are those between microorganisms and leachate molecules on the soil surfaces and their effects on microbial catabolism. These interactions must be understood so that the effects of various soil types on, and in, the landfill can be predicted. Soils differentially adsorb organic and inorganic compounds and microorganisms. Because of this, soil has, potentially, a great buffering capacity so that the landfill as a bioreactor may be protected against surges of certain chemical or even physical challenges.

In addition to their roles in landfills, there are several important reasons for studying the interactions between soil, microorganisms, and pollutants (SMPs). Pollutants in the soil environment often migrate to the groundwater or open water bodies. Understanding the mechanisms of SMP interactions will, thus, facilitate risk assessment in various compromised ecosystems and will be of importance for environmental impact assessments. This understanding may be applied to potential pollutants, including agricultural and industrial chemicals.[4] Bioreclamation of polluted soils is greatly dependent on the characteristics of SMPs as well as their interactions.[5] Pesticides added to soil are subject to microbial catabolism as well as adsorption by soil. These interactions, therefore, affect pesticide activity and environmental safety.[6,7]

A. THE IMPORTANCE OF ADHESION

The soil components provide surface area for interaction with microorganisms and pollutants.[8,9] Although soil coverings (intermediate and final) constitute only a small fraction of the landfill volume, the total surface area per unit volume of the soil greatly exceeds that of refuse. It is, therefore, probable that the soil has a significant influence on both microorganisms and pollutants migrating through the refuse mass. Upward migration of leachate, by capillary rise, takes place mainly through the micropores of the soil although vertical migration may also be via saturated landfill gas. The surface area (surrounding the pore) to pore space ratio is much higher than the average for the pores which would normally be involved in downward (gravitational) migration of leachate. Adhesion of microorganisms to surfaces has been found to influence metabolic activity and, therefore, also catabolic capabilities in many instances.[10-12] Adhesion may also afford greater resistance/protection of microorganisms to toxic substances.[13] This is probably the most important factor in considering the role of soil in landfills. Research indicates that not only the microorganisms but also their enzyme activities are influenced by adhesion to different soil components.[14] The adhesion of microorganisms, together with adsorption of organic molecules to soil components, is particularly important in pollutant attenuation in soil.[15]

II. SOIL COMPONENTS AND THEIR PROPERTIES

To understand the interactions among SMPs, a brief discussion of each is warranted with specific reference to adhesion/adsorption.

A. CLAYS

The nature of clays is such that they differ in their adsorption capacities (cation and anion) and specific surface areas.[16-18] Unfortunately, the effects of these properties on microorganism and pollutant interactions are not well understood. In the silicate clay minerals the characteristic net negative charge originates mainly from isomorphous substitution of Al^{3+} for Si^{4+} in the tetrahedral sheets, and Mg^{2+}, and other divalent cations, for Al^{3+} in the octahedral sheets.[19-21] The negative charge is balanced by adsorbed cations so that the clays are electrically neutral.

Some clay-sized (<2 µm) components in soils possess a net positive electrical charge which is balanced by anions so that the system as a whole is again electrically neutral (anion exchange capacity). Anion exchange capacity (AEC) is due to protonation of OH^- groups on the broken edges of clays and on oxide and hydroxide mineral surfaces. The AEC of soils may have a particular influence on soil microorganisms and chemicals with negative surface charge.[22] Anion exchange capacity is more prone to fluctuation due to pH changes than cation exchange capacity (CEC) since CEC is mainly due to the permanent negative charge of most clays and is also greatly dependent on soil mineral composition.[19,22] The CEC:AEC ratio is probably just as important in adhesion of microorganisms to soil, as CEC or AEC alone. The CEC:AEC ratio determines the relative net negativity of different clays and the ability of negatively charged organisms and molecules to be attracted or repulsed by these clays.[22]

B. OXIDES AND HYDROXIDES

Together with hydrous aluminosilicates, many soils also contain other oxides, oxyhydroxides, and hydroxides of Fe^{3+}, Al^{3+}, Mn^{4+}, Ti^{4+}, and Si^{4+}. Only the iron and aluminum oxide minerals will be considered here since they constitute the major reactive oxide components. The basic characteristics of the aluminum hydroxides, oxyhydroxides, and iron hydroxides were reported by Hsu[23] and Schwertmann and Taylor.[24] Unlike silicate clays, where the inherent charge is mostly permanent because of isomorphous substitution, the exchange capacity of the oxide minerals is pH dependent.[19] Oxide minerals at their isoelectric point or zero point of charge (ZPC) are neutral. Increasing pH favors reaction with OH^- and creation of CEC. By contrast, lowering pH favors reaction with H^+ and creation of AEC. Leachate, which is often acidic, may thus increase the AEC which, in turn, may increase initial adhesion of microorganisms to these surfaces. In general, microorganisms have negative surface charges at pH values above 3. The soil oxide minerals often form coatings on the surfaces of other soil minerals such as silicate clays and sand-sized particles. These coatings alter the surface characteristics of the minerals[25] and may influence microbial and organic compound interactions with these minerals. The oxide minerals have been found to be of importance in landfill leachate attenuation.[26] Although the information on microbial interactions with soil oxides and hydroxides is scarce, workers have reported interactions of soil oxides with organic compounds[27] and microbially produced enzymes.[14,28] These reports are mainly of sorption and crystallization reactions with the oxides and hydroxides.

C. ORGANIC MATTER

Humus fractionation studies play an increasingly important role in research into the behavior of pollutant chemicals in the soil environment.[29,30] It has been found that the binding of organic pollutants to water-soluble humic substances increases both their mobility in soil and their leaching,[31] whereas binding to high molecular weight substances results in nonextractable residues.[32] Biodegradation and bioavailability of nonextractable soil-bound residues are usually greatly reduced compared with the free compounds. Mahro and Kästner[33] found polycyclic aromatic hydrocarbons (PAHs) to be incorporated into humus polymers via oxidative coupling. This incorporation reduced the extractability and biodegradation of the compounds. Such incorporation or complexation is believed to be of benefit since the PAHs are not desorbed. Unlike adsorption sites on charged surfaces, which can be saturated, this form of incorporation does not reach saturation but increases the adsorption sites for other pollutants because of the hydrophobic nature and CEC (due to numerous functional groups) of the humus. Rebhun et al.[34] studied the sorption/partitioning of several organic contaminants with a wide range of hydrophobicities on clay-humic complexes. The adsorption constants of the humic fraction were found to be 8 to 20 times higher than those of the "pure" clays used. In soils with low to medium organic matter contents (0.5 to 1.5%), however, the contribution of the clay minerals to adsorption was shown to be quite significant, in spite of the fact that half of the sorption sites on the mineral surfaces were blocked by humic substances.

The less refractile components of soil organic matter (SOM) may also serve as carbon sources for microorganisms growing in the soil. The implications of having readily degradable SOM in an SMP environment are that it may be used as a preferential carbon source so that organic pollutant catabolism does not occur. This is of particular importance where the pollutant molecules are adsorbed on the soil (or SOM) and the alternative carbon source is more soluble and available for microbial catabolism.

The SOM is usually concentrated in the surface horizon. Soils used in landfills are likely to, initially, have very low amounts of organic matter since these soils will have been excavated and will, for the most part, be subsoil. The formation of humus and incorporation of pollutants into the humus-polymers by oxidative complexing is only able to occur in the presence of oxygen.[33] This process could, thus, only occur either in the early stage of landfilling, when enough oxygen is present, or in final covering soils exposed to oxygen. The importance of humus as an absorbent for pollutant chemicals should also be kept in mind when vegetating the completed landfill. Some plant species such as grasses accelerate humus formation in soil to a much greater extent than, for example, trees. The greater humus content of the soil would, therefore, be beneficial in attenuating vertically migrating leachate chemicals and, possibly, gases.

SOM also affects the soil's physical characteristics. Even when conditions are not conducive to humus formation the organic content of the soil in the landfill will increase with time as organics from refuse leachate accumulate on soil particles and result in biofilm development. Bulk density, porosity, and aeration, as well as water and heat movement, are all influenced by this accumulation.[35] SOM has a direct effect on soil structure and aggregation[36] but because of the relatively stable water content and subsequent absence of wetting and drying cycles it is uncertain whether soil structure formation would occur in soil within the refuse mass. The significance of soil structure and aggregation to SMP systems is discussed later.

III. POLLUTANTS

Pollutant characteristics of importance to SMP systems are recalcitrance, adsorption properties, and solubility. Some of these are discussed below.

A. METALS

Many studies have been made to determine the effects of soil and soil properties on migrating metal pollutants[37,38] as well as their behavior in landfills.[39,40] Metals, unlike organic chemicals, are not catabolized by microorganisms and are, therefore, not removed from the environment in which they occur. Although the interactions of the organic components of leachate are the main factors considered in this chapter, the effects of heavy metals cannot be ignored. Campbell et al.[41] found that, under intensive leaching with a mixed metal/carboxylic acid solution, microbial colonization of the unsaturated zone was inhibited by the presence of heavy metals (Ni and Cd). It is, however, difficult to know whether microbial resistance would have been less if the microorganisms had been unattached. Soil surfaces can adsorb heavy metal ions and, thus, increase metal concentration at the soil-water interface where the high concentrations may be detrimental to surface-attached bacteria. The increased surface sorption of the metal ions to soil components may, however, also reduce the concentrations of the metals in solution to subcritical levels. Finally, microorganisms may also be protected against possible toxicity due to surface attachment.[13] Here it seems as if microorganisms associated with surfaces (soil and refuse) are able to withstand much higher concentrations of free heavy metals than is the case with free-living species. Humic acids have also been found to significantly decrease the adsorption of heavy metals to soils. This is possibly due to the strong complexing ligands of the humic acids competing with the surface functional groups of soil for the free heavy metals in the aqueous phase. These effects increase both mobility and bioavailability.

B. PARTITIONING OF ORGANICS

The vast array of organic landfill leachate components[42-45] complicates examination of the SMP system. The interactions of soil and pollutants (and thus also bioavailability) are related to the partitioning of the chemical compounds between the liquid and solid phases in the soil, as well as to the kinetics of desorption.[29,46-48]

According to Green et al.[49] the octanol/water partitioning coefficient is probably the single most important laboratory-determined parameter for predicting the movement and adhesion of organic compounds in soils. This parameter can, unfortunately, only be measured over long periods of time (of the order of a month). The coefficient of permeability (K) can, however, be related to more easily measured parameters of soils and organics:

$$K = QL/AH$$

where

> K = coefficient of permeability (cm/s)
> Q = the flow of the percolate (ml/s)
> L = the cross-sectional area of the sample (cm^2)
> H = the average head of the fluid medium on the sample (cm)

Green et al.[49] found that the hydrophobic or hydrophilic nature of the organic compounds, as measured by the octanol/water partitioning coefficient (or, approximately, by the dielectric constant), were important for predicting the solvent's rate of flow through soils. The octanol/water partitioning coefficient measures the tendency of molecules to escape from the aqueous phase. Hydrophobic substances, such as benzene, xylene, and carbon tetrachloride (log octanol/water partitioning coefficients of 2.13, 3.15, and 2.64, respectively; water = −1.15), would be expected to adhere more strongly to the soil solid phase than water. The octanol/water partitioning coefficient was found to be more important than the organic compound density and viscosity in predicting flow rate (and adhesion) through soil.[49]

An important principle here is that, for hydrophobic compounds which are often present in concentrations exceeding their maximum water solubility threshold, adsorption is not adsorbent (soil or refuse) dependent.[3] The adsorption to soil is, thus, due to the saturation of the aqueous phase. When the maximum water solubility concentration of the compound has not been exceeded, the adsorption is controlled by the adsorbent; thus, a high SOM content (hydrophobic surfaces) should result in increased adsorption of hydrophobic molecules.

C. BIODEGRADABILITY

Recalcitrance of organics to microbial catabolism is often related to the substitution of different groups on the molecule.[50,51] Substitution with halogens is particularly important. Not only the identity of the halogen but also the position of substitution have been found to be important.[50] The different capacities for biodegradation of selected halogenated aliphatics have also been found to differ both under different redox potentials and with the carbon chain length.[51] Different substitutions and functional groups also affect the charge and polarity of the molecule and, therefore, the adsorption to soil. The biodegradation properties of the pollutants will, thus, be significantly influenced by adhesion to soil components.[52,53]

Under aerobic conditions, in laboratory tests, it has been shown that substituted phenols not only sorbed irreversibly to clays and soils but also were transformed into polymerized moieties.[44] This suggests that substituted phenols will leach through soil more readily under the anaerobic conditions of the landfill and that such migrations may be inhibited under aerobic conditions. This may also be the case for other compounds. This phenomenon is probably related to the polymerization and oxidative coupling of polyaromatic hydrocarbons in humus complexes.[33]

IV. MICROORGANISMS

A. BACTERIAL ADHESION

The importance of adhesion in the SMP system has been discussed above. The mechanisms of general microbial adhesion (rather than adhesion to soil surfaces specifically) will be discussed here. Although the information is of a fundamental nature, it may be extrapolated to microbial adhesion in the SMP system.

Attachment can be divided into two steps. First, organisms adhere in a process that is governed by the physicochemical surface properties of the bacteria and the solid and the type of solute.[54] In the second step, microorganisms anchor themselves to a surface using specific appendages or cell surface structures. The process depends on the type of bacterium in combination with the type of surface. The adhesion step can be divided into two separate stages: reversible and irreversible adhesion. Reversible adhesion is similar to the deposition of a bacterium on a surface so that two-dimensional Brownian motion can still take place and the bacterium can be removed from the surface. Irreversible adhesion is where bacteria can no longer be removed from the surface and so are incapable of exhibiting Brownian motion. Bacterial adhesion is a complicated process because cells are not "ideal" particles. Furthermore, internal chemical reactions can lead to changes in molecular composition, both intercellularly and at the surface of the bacterium. These chemical processes continue after adhesion. Attached cells are, therefore, rarely in complete physicochemical equilibrium with their environment.

The total long-range interaction between bacteria (living colloidal particles) and a surface over a distance of more than 1 nm is, according to the Derjaguin, Landau, Verwey, and Overbeek (DLVO) theory, a summation of Van der Waals and Coulombic interactions.[54] Generally, surfaces of particles and microorganisms are charged. Because of electroneutrality in water, the charge on the surface is neutralized by a countercharge which is diffusely distributed around the particle. The thickness of the diffuse double layer is a function of the ion charge and ion concentration. The Gibbs free

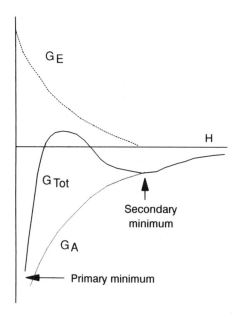

Figure 1 Gibbs energy of interaction between a sphere and a flat surface with the same charge sign (G_{Tot}), according to the DLVO theory. G_A = free energy of the Van der Waals forces; G_E = free energy of the electrostatic interaction; and H = separation distance.

energy (G_E) of the electrostatic interaction is determined by the electroki-netic (or zeta) potential of the surfaces. As stated above, most natural surfaces and bacteria are negatively charged.[55] Figure 1 shows a charac-teristic plot of the total interaction Gibbs energy (G_{tot}, which is a summation of G_A and G_E) as a function of the separation (H) between a bacterium and a negatively charged surface and shows two minima. If a bacterium reaches the primary minimum, short-range forces dominate the adhesive interaction and the DLVO theory cannot be used to predict the interaction energy. The secondary minimum is never greatly negative so that particles captured in this minimum generally show reversible adhesion. When bacteria make direct contact with a surface (separation distance H = 0) the interaction energy can be calculated from the assumption that the interfaces between solid/liquid and bacterium/liquid are replaced by a solid/bacterium inter-face. If the Gibbs free energy for adhesion is negative the adhesion is thermodynamically favored and will proceed spontaneously. Energy values below 4×10^{-20}J/cell result in irreversible adhesion (primary minimum) which is much stronger than adhesion in the secondary minimum (Figure 1). Short-range interactions can, however, only become effective when long-range interactions allow a particle to approach a surface. A high maximum in G_{tot} would prevent such an approach.

The hydrophobicity of a compound is an indication of its tendency not to interact with water. Hydrophobic interactions result from the fact that water-water contacts are thermodynamically more favorable than contacts between two nonpolar groups or between a nonpolar group and water.[54] Nonpolar groups tend to be rejected from aqueous medium. In a "normal" soil system, this should effectively mean that hydrophobic microorganisms should adhere to the soil and organic matter particles rather than staying in solution. Generally, the Gibbs free energy maximum decreases with increasing hydrophobicity and results in higher adhesion strength.

There is agreement between experimental observations and theory that primary minimum adhesion (irreversible) is to be expected in the case of very strong Van der Waals attraction (i.e., both surfaces are hydrophobic) or when electrostatic forces are attractive or only weakly repulsive.[54] Normally, however, adhesion is found to be reversible. This indicates that, in thermodynamic terms, the interaction between a bacterium and a surface is relatively weak (Gibbs free energy $>4 \times 10^{-20}$J/cell). Bacteria become more hydrophobic and show increased adhesion during the exponential growth phase and at high dilution rates in a chemostat. The reason for this phenomenon is not yet known.[54,56] The cause of cell wall hydrophobicity is also not totally clear but it appears that thin fimbriae and fibrils play an important role as hydrophobins.[57-60] Bacteria, initially adhering in the secondary minimum may, with time, simply pass through the energy barrier (if it is not too high) or penetrate the energy barrier by protruding fibrils or fimbriae. The electrostatic repulsion energy depends more strongly on the particle radius than the Van der Waals forces. Fimbriae (because of their small radii) can readily adhere in the primary minimum and thus bridge the gap between surface and bacterium. There are indications that surface polymers may sterically hinder the close approach of two surfaces and force the particles to adhere in the secondary minimum. The occurrence of secondary minimum adhesion is, thus, not necessarily due to electrostatic repulsion. Most of the work on hydrophobic adhesion has been done in simulated studies rather than *in situ*. Furthermore, the soil environment is often a relatively low nutrient environment so that starvation responses occur which could greatly alter hydrophobicity due to the possible loss of fimbriae and many other factors. Hydrophobic adhesion has also been shown to be influenced by both chromosomal and plasmid control[61] and is not only dependent on the hydrophobicity of the bacterial cells but also on that of the adsorbent surface.[62] Soil components differ in their hydrophobicity. The SOM especially has been shown to be more hydrophobic than the mineral components.[36] This may be very important for microbial adhesion since it should reduce the maximum Gibbs energy barrier and facilitate adhesion. All of the interactions discussed here are of importance in landfill soils because of the effect of adsorption (attachment) and its role in protecting microorganisms against potentially toxic chemicals. These interactions are also important in microbial migration.

B. COSOLVENCY

Another interesting aspect of hydrophobic interactions (and the octanol/ water partitioning concept) is that hydrophobic interaction reversal may occur when the soil solution contains not only water but also a nonpolar cosolvent(s), as may occur in landfills. This reversal affects hydrophobic molecules[3,63] as well as hydrophobic microorganisms. Cosolvency results in a lowering of the polarity of water which allows desorption of hydrophobic entities. This is of particular importance when co-disposal is employed which could result in cosolvency. Through desorption of these compounds, the leachate concentrations of molecules previously adsorbed due to hydrophobicity will increase. Sorption and low solubility are major factors preventing biodegradation of hydrophobic molecules.[64,65] Cosolvency, therefore, not only increases the soluble concentration but also increases biodegradation of hydrophobic molecules. Leachate quality is affected by the ratio of desorption to biodegradation (as will be discussed later). When practicing co-disposal with cosolvents it should, however, be kept in mind that most cosolvents are toxic to microorganisms at elevated concentrations. These toxicity threshold concentrations have been found to be much higher in the presence of soil than in soil-free systems. The tolerance of elevated concentrations in the presence of soil cannot be accounted for by adsorption of cosolvents to soils and is most probably because of the microbial interactions and protection at soil surfaces.

C. BIOFILMS

Biofilm development and its importance have been well documented.[66] Biofilm accumulation within porous media such as soil can substantially reduce the hydraulic conductivity.[67] The increased resistance to flow is due to the reduction of the effective pore space caused by attached cells and their extracellular matrices. Microorganisms growing as biofilms in the subsurface (or soils) have an advantage over suspended species in that they can remain near the source of fresh substrate and nutrients contained in the groundwater which flows by them. The rate of biofilm growth and, thus, the rate of biotransformation is, therefore, strongly influenced by transport characteristics, including velocity distribution within pores, dispersivity, surface roughness (all of which are dependent on soil composition), molecular diffusivity, and other variables which affect the delivery rate of substrate and nutrients to the growing cells. Hydraulic flow rate has an influence on retention time which, in turn, has an influence on biotransformations and biodegradation of organic pollutant chemicals. The longer an organic molecule is retained within a soil pore, the greater the probability that the molecule will be adsorbed or catabolized. The interactions between hydraulic conductivity and adsorption and catabolism are discussed later. Biofilm development *per se* also seems to be beneficial because of accumulation (organics and inorganics) and subsequent degradation (organics) of leachate pollutants.

D. MICROBIAL INTERACTIONS

Biodegradation studies are complicated by microbial interactions. The interactions are varied and depend on environmental circumstances.[68,69] In general, inter-, and intraspecies interactions are poorly understood and little is known of the physiological changes that occur after adsorption.

E. PHYSICOCHEMICAL INTERACTIONS OF ADHESION

As discussed earlier, the properties of clay minerals which affect surface interactions with microorganisms have not been closely studied but attempts have been made to identify the properties that affect the adsorption of soluble organic materials.[70] Some soluble organic compounds are similar to components of the surface structures of microbial cells and, therefore, the properties of these compounds may be relevant to adhesion of microorganisms. Before making extrapolations, however, it should be noted that free soluble organic compounds are not confined to the more rigid structures as is the case with compounds in cell surfaces. Hydrophobicity, which could also be included under this heading, has already been discussed.

F. CHARGE INTERACTIONS

The cations on the clay surfaces and their interactions with microorganisms have been well documented.[71-74] In general, it has been shown that, if a net positively charged biological entity (an organic molecule, particle, or microorganism) is involved, surface interactions are usually greater when the charge compensating cations have a low valency (i.e., monovalent). The most probable reason for this phenomenon is that replacement of monovalent cations is easier than that of multivalent cations.[75]

Most biological cells have a net negative charge under normal soil conditions. Surface interactions between these and clays are usually greater when the valence and the ionic strength of the exchangeable cation solution are greater. This is probably because of a decrease in the thickness of the diffuse double layer[54] which effects a decrease in the Gibbs maximum energy barrier (Figure 1). This enables the clays and the negative biological entities to approach each other more closely so that H-bonding and Van der Waals interactions become effective and adsorption is favored. The

ionic strength of landfill leachate, thus, has a pronounced effect on microbial adhesion and migration in soil. The above process, called the Schulze-
Hardy rule, must be distinguished from the process of cation bridging where
multivalent exchangeable cations on the surfaces of the clays act as bridges
between the clay and the negatively charged biological entity.[22] Cation
bridging is not, however, a valid theory since the cations would be hydrated
and so adhesion would be the indirect result of protonation or water
bridging. Multivalent cations also have the capacity to reduce the expan-
sibility of swelling 2:1 layer clays which reduces their total surface area.
This can result in the reduction of macromolecule (protein) binding.[75,76]

The net negativity of biological cells is dependent on the pK_a values of
the dissociable functional groups and the ambient pH.[22,75] The pI (pH at
the isoelectric point) of the entity, which is an empirical summation of the
pK_a values of all components capable of accepting or releasing protons and
the pH of the ambient solution, determines the net charge of the entity. It
is apparent that landfill leachate could enforce an unnatural pH condition
on the soil. Adhesion characteristics of the microorganisms may, therefore,
be different under such conditions. This may be particularly significant for
relatively small molecules since pI values are only important for such molecules. Adhesion of large molecules and particulate entities (and microorganisms) are not as dependent on pI values since they may have positively
charged sites even at pH values above their pI when their net charge is negative.

Some heavy metals such as Cd, Cu, Cr, Ni, and Zn, which have been
found in landfill leachate,[40] have been shown to cause reversal in the charge
of bacterial cells (and clays) in monocultures to positive values, at pH levels
above their pI values.[77] This may be important in the landfill environment
where heavy metals may be present and, thus, cause increased attachment
of microorganisms to soil particles because of the charge reversal. Attachment, as discussed above, may significantly influence microbial catabolism
of pollutant compounds.

G. PROTONATION

Protonation seems to be the most important feature of interactions between clays and biological cells. Protons from the clay surfaces or, more
likely, from the associated water, are transferred to the biological cell. This
transfer usually takes place to O and N and, to a lesser extent, S and P.
This protonation adds a positive charge to the biological entity so that it
may become less negative (or neutral, or positively charged) and surface
interaction with the clay can then occur via H-bonding, Van der Waals
forces, or Coulombic mechanisms.[22] Protonation is a function of surface
pH (pH_s) and the pH of the adjacent bulk solution (pH_b). The pH_b is usually
higher than the pH_s due to the formation of the diffuse double layer around

the clay. The difference between pH_s and pH_b (ΔpH) depends on the charge-compensating cations. The higher the concentration of basic charge-compensating cations, the fewer protons on the exchange surface and, thus, the higher the pH_s. The relatively low pH and high cationic load of leachate migrating through the soil could affect the ΔpH. Although the pH_b may be above the pI value of the bacteria, the pH_s may be below it so that bacteria close to the surfaces of the clays may be protonated, thereby increasing their interactions with the clay surfaces. The pH_s may also be below the pK_a value of organic chemicals, although the pH_b might be above. This could facilitate cationic adsorption of the organic compound to the surface.[29] Organic conditioning films may also be important in adhesion and act as intermediates in H-bonding between the cells and clays.

V. PARTICLE AGGREGATION, HYDRAULIC CONDUCTIVITY, AND MICROBIAL CATABOLISM

There is probably a correlation between hydraulic conductivity and microbial catabolism of both adsorbed and soluble compounds.[3] In fermentation studies, the important factor of dilution rate is a measure of the medium flow rate into the bioreactor divided by its volume. This is, in some cases, assumed to be applicable to soil systems. The dilution rate gives an indication of the rate at which medium should be applied to a vessel (or the pore space occupied by soil microorganisms) in order to facilitate optimum microbial growth. If this flow is too fast it leads to washout of the microorganisms and if it is too slow it leads to starvation unless the concentration of the medium is sufficient to satisfy the maintenance energy requirement of the microorganisms. In the soil system, however, there are several complicating factors. The first problem in using dilution rates to understand the dynamics of microbial degradation of pollutants is that the soil volume occupied by microorganisms is difficult to determine. Under saturated conditions this volume approximates to the total pore space. A further problem here is that the surface area for attachment, relative to the pore volume, is much greater than in other reactors. Under unsaturated conditions an estimation of the volume occupied by the microorganisms becomes even more difficult since it cannot simply be determined by calculating or measuring the soil water content. The soil water closely associated with the soil surfaces does not behave as free water and assumes the character of crystalline water.[19] This type of water, relative to the free water, increases with a decrease in the water content. Crystalline water also causes a sharp decline in the water activity which, in turn, also effects a decline in microbial activity.[22]

To complicate the problem of understanding the dynamics of the SMP system there is also the factor of adsorption of biodegradable compounds

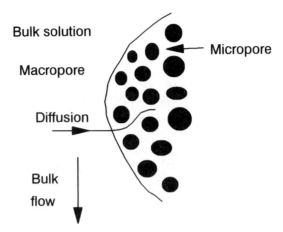

Figure 2 Diagrammatic representation of a soil aggregate, made up of several soil particles and showing micro- and macropores in relation to liquid migration.

(due to hydrophobic or charge interactions). Because of the equilibrium between adsorbed and solution concentrations of particular compounds the hydraulic conductivity can greatly influence microbial catabolism. Figure 2 is a diagrammatic representation of a soil aggregate which is made up of several soil particles. The spaces between these aggregates are mainly macropores and facilitate mass flow (bulk flow). Microbial catabolism mostly occurs at the surface of the aggregate or in the micropores if the bulk flow (related to hydraulic conductivity) is such that microbial washout occurs. The macropores are usually drained at field capacity so that microbial activity is restricted to the micropores. Replenishment of the micropores with nutrients or pollutants only occurs if there is sufficient time for diffusion between the bulk solution in the macropores and the micropores. Because of the slight solubility of most hydrophobic molecules, diffusion also occurs from the micropores, where most of the compounds are adsorbed, to the bulk solution which probably, initially, has a lower concentration. If the hydraulic conductivity is low (or zero) the bulk solution will soon reach maximum solubility concentration so that no more diffusion out of the micropores into the bulk solution occurs. If, however, bulk flow does occur to an appreciable extent the bulk solution will be constantly below the maximum solubility concentration so that the concentration gradient between micro- and macropores will stay intact, thus sustaining diffusion. Because of this effect on the concentration in the bulk solution the microbial activity is also affected. The extent to which migration of particularly hydrophobic pollutants occurs in soil (and, therefore, affects microbial catabolism) depends greatly on the diffusion distance between the micropores in the aggregate and the macropores in which most of the bulk flow occurs.[3,78]

This phenomenon is greatly affected by soil structure, both macro-soil aggregates and micro-aggregates. These micro-aggregates may also be present in relatively homogeneous soil. All of these intricate variables make the interpretation of soil column (microcosm) studies very difficult and thus justifies a combination of soil columns and soil batch studies to determine the interactions of the SMP system at soil surfaces.

A. ADSORPTION ISOTHERMS

On the whole, soil has a great buffering capacity for hydrophobic contaminants and charged molecules. Soil is, thus, able to reduce the environmental risk of migrating adsorbable pollutants until biodegradation (if possible) occurs. This is particularly important in landfills where such pollutants (organic or inorganic) may be potentially toxic at high aqueous concentrations. The soil could, thus, act as a buffer to lower the soluble concentration to below the toxicity threshold. This would also allow more time for microbial adaptation to occur. This capacity is of great importance where landfills are viewed not only as dumps but, increasingly, as bioreactors which facilitate breakdown of compounds to exploitable products. It should, however, also be recognized that every soil type has a finite buffering (adsorptive) capacity which could be exceeded. At that point it would no longer adsorb any of the challenging molecules, thus, an increase in equilibrium aqueous concentration would not result in increased adsorption.[3] This point of saturation can, however, be determined with adsorption isotherms. With most hydrophobic compounds, the solubility of the molecule is the most limiting factor and not the soil adsorption capacity. Increased adsorption occurs not only because of the sorption capacity of the soil but also because of the hydrophobic nature of the compound. These compounds are repelled from the aqueous phase and, thus, adsorb to soil. In this case, the adsorption isotherm would show that an increase in the adsorbed molecule concentration does not increase the equilibrium aqueous phase concentration. As the compounds are catabolized, usually from the aqueous phase, the soil-adsorbed concentration replenishes the soluble concentration so that it is kept at its solubility maximum. This continues until the adsorbed amount is insufficient to sustain the solubility maximum. Water (leachate) analysis, therefore, gives only limited information about the state of contamination and should be carefully interpreted.

VI. CONCLUSIONS

The soil has a great buffering capacity and is able to adsorb potentially toxic chemicals, thus reducing the effective aqueous phase concentration of each compound. Soil also seems to have a significant effect on the microbial populations by affording protection against chemical challenges, probably by attachment to surfaces. From a physicochemical point of view, soil *per se* seems to be beneficial for the landfill as a whole. Thus, the soil layers (cappings) should protect both microbial populations and, ultimately, covering vegetation against chemical challenges. With respect to this both the clay (content and type) and organic matter contents of the soil are important variables. Although it may not always be possible to use a specific soil type for landfill purposes it is nevertheless important to understand the effects of soil type on the SMP system. This is probably even more important in environmental clean-up and risk assessment and should under no circumstances be ignored. The importance of the *in situ* soil physical conditions should never be assumed to override the physicochemical SMP effects at the soil surfaces.

REFERENCES

1. Smith, K. A., Bremner, J. M., and Tabatabai, M. A., Sorption of gaseous atmospheric pollutants by soils, *Soil Sci.,* 116, 313, 1973.
2. Metcalfe, D. E. and Farquhar, G. J., Modelling gas migration through unsaturated soils for waste disposal sites, *Water Air Soil Pollut.,* 32, 247, 1987.
3. Knox, R. C., Sabatini, D. A., and Canter, L. W., *Subsurface Transport and Fate Processes,* Lewis Publishers, London, 1993.
4. Knezovich, J. P., Hirabaya, J. M., Bishop, D. J., and Harrison, F. L., The influence of different soil types on the fate of phenol and its biodegradation products, *Chemosphere,* 17, 2199, 1988.
5. Piotrowski, M. R., Bioremediation of hydrocarbon contaminated surface water, groundwater and soils: the microbial ecology approach, in *Hydrocarbon Contaminated Soils and Groundwater,* Vol. 1, Kostecki, P. T. and Calabrese, E. J., Eds., Lewis Publishers, Chelsea, MI, 1991, 203.
6. Boesten, J. J. T. I. and van der Linden, A. M. A., Modelling the influence of sorption of and transformation on pesticide leaching and persistence, *J. Environ. Qual.,* 20, 425, 1991.
7. Boesten, J. J. T. I., van der Pas, L. J. T., Smelt, J. H., and Leistra, M., Transformation rate of methyl isothiocyanate and 1,3-dichloropropene in water-saturated sandy subsoils, *Neth. J. Agric. Sci.,* 39, 179, 1991.
8. Scott, H. D., Wolf, D. C., and Lavy, T. L., Apparent adsorption and microbial degradation of phenol by soil, *J. Environ. Qual.,* 11, 107, 1982.
9. Fontaine, D. D., Lehmann, R. G., and Miller, J. R., Soil adsorption of neutral and anionic forms of a sulphonamide herbicide, flumetsulam, *J. Environ. Qual.,* 20, 759, 1991.
10. Fletcher, M., Effect of solid surfaces on the activity of attached bacteria, in *Bacterial Adhesion, Mechanisms and Physiological Significance,* Savage, D. C. and Fletcher, M., Eds., Plenum Press, New York, 1985, 339.

11. Bar-Or, Y., The effect of adhesion on survival and growth of microorganisms, *Experientia,* 46, 823, 1990.

12. van Loosdrecht, M. C. M., Lyklema, J., Norde, W., and Zehnder, A. J. B., Influence of interfaces on microbial activity, *Microbiol. Rev.,* 54, 75, 1990.

13. Kefford, B., Kjelleberg, S., and Marshall, K. C., Bacterial scavenging: utilization of fatty acids localized at a solid-liquid interface, *Arch. Microbiol.,* 133, 257, 1982.

14. Gianfreda, L., Rao, M. A., and Violante, A., Adsorption, activity and kinetic properties of urease on montmorillonite, aluminium hydroxide and Al(OH)$_x$-montmorillonite complexes, *Soil Biol. Biochem.,* 24, 51, 1992.

15. Giles, C. H., Smith, D., and Huitson, A., A general treatment and classification of the solute adsorption isotherm, in *Adsorption Phenomena,* Harter, R. D., Ed., Van Nostrand Reinhold, New York, 1986, 323.

16. Harter, R. D., Reactions of minerals with organic compounds in the soil, in *Minerals in Soil Environments,* 2nd ed., Dixon, J. B. and Weed, S. B., Eds., Soil Science Society of America, Madison, WI, 1989, 709.

17. Harter, R. D., *Adsorption Phenomena,* Van Nostrand Reinhold, New York, 1986.

18. Burns, R. G., Interaction of microorganisms, their substrates and their products with soil surfaces, in *Adhesion of Microorganisms to Surfaces,* Ellwood, D. C., Melling, J., and Rutter, P., Eds., Special Publication of the Society for General Microbiology, Academic Press, New York, 1979, 193.

19. Foth, H. D., *Fundamentals of Soil Science,* 7th ed., John Wiley & Sons, New York, 1984.

20. Dixon, J. B., Kaolinite and serpentine group minerals, in *Minerals in Soil Environments,* 2nd ed., Dixon, J. B. and Weed, S. B., Eds., Soil Science Society of America, Madison, WI, 1989, 467.

21. Borchardt, G. A., Smectites, in *Minerals in Soil Environments,* 2nd ed., Dixon, J. B. and Weed, S. B., Eds., Soil Science Society of America, Madison, WI, 1989, 675.

22. Stotzky, G., Mechanisms of adhesion to clays, with reference to soil systems in *Bacterial Adhesion, Mechanisms and Physiological Significance,* Savage, D. C. and Fletcher, M., Eds., Plenum Press, New York, 1985, 305.

23. Hsu, P. H., Aluminum oxides and oxyhydroxides, in *Minerals in Soil Environments,* 2nd ed., Dixon, J. B. and Weed, S. B., Eds., Soil Science Society of America, Madison, WI, 1989, 331.

24. Schwertmann, U. and Taylor, R., Iron oxides, in *Minerals in Soil Environments,* 2nd ed., Dixon, J. B. and Weed, S. B., Eds., Soil Science Society of America, Madison, WI, 1989, 379.

25. Hendershot, W. H. and Lavkulich, L. M., Effect of sesquioxide coatings on surface charge of standard mineral and soil samples, *Soil Sci. Soc. Am. J.,* 47, 1252, 1983.

26. Lucas, J. L. and White, C. C., Design and instrumentation of semi-permeable attenuation blanket beneath a domestic waste landfill, in Proceedings: International Land Reclamation Conference, Grays, England, Industrial Seminars, Tunbridge Wells, 1983, 299.

27. Sibanda, H. M. and Young, S. D., Competitive adsorption of humus acids

and phosphate on goethite, gibbsite and two tropical soils, *J. Soil Sci.*, 37, 197, 1986.

28. Huang, P. M. and Violante, A., Influence of organic acids on crystallization and surface properties of precipitation products of aluminum, in *Interactions of Soil Minerals with Natural Organics and Microbes*, Soil Science Society of America Special Publication No. 17, Huang, P. M. and Schnitzer, M., Eds., Madison, WI, 1986, 159.

29. Senesi, N. and Chen, Y., Interactions of toxic chemicals with humic substances, in *Toxic Organic Chemicals in Porous Media*, Gerstl, Z., Chen, Y., Mingelgrin, U., and Yaron, B., Eds., Springer-Verlag, Berlin, 1989, 37.

30. Andreux, F., Portal, J. M., Schiavon, M., Bertin, G., and Barriuso, E., The usefulness of humus fractionation methods in studies about the behaviour of pollutants in soils, *Toxicol. Environ. Chem.*, 31/32, 29, 1991.

31. Lafrance, P., Banton, O., Campbell, P. G. C., and Villeneuve, J. P., A complexation-adsorption model describing the influence of dissolved organic matter on the mobility of hydrophobic compounds in groundwater, *Water Sci. Technol.*, 22, 15, 1990.

32. Scheunert, I., Mansour, M., and Andreux, F., Binding of organic pollutants to soil organic matter, *Int. J. Environ. Anal. Chem.*, 46, 189, 1992.

33. Mahro, B. and Kästner, M., Mechanisms of microbial degradation of polycyclic aromatic hydrocarbons (PAH) in soil-compost mixtures, in *Contaminated Soil '93*, Arendt, F., Annokkèe, G. J., and van der Brink, W. J., Eds., Kluwer Academic Publishers, The Netherlands, 1993, 1249.

34. Rebhun, M., Kalabo, R., Grossman, L., Manka, J., and Rav-Acha, C., Sorption of organics on clay and synthetic humic-clay complexes simulating aquifer processes, *Water Res.*, 26, 79, 1992.

35. Marshall, T. J. and Holmes, J. W., *Soil Physics*, 2nd ed., Cambridge University Press, Cambridge, 1988.

36. Sullivan, L. A., Soil organic matter, air encapsulation and water-stable aggregation, *J. Soil Sci.*, 41, 529, 1990.

37. Fuller, W. H., Investigation of Landfill Leachate Pollutant Attenuation by Soils, U.S. EPA-600/2-78-158, Washington, DC, 1978.

38. Fuller, W. H., Soil modification to minimize movements of pollutants from solid waste operations, *Crit. Rev. Environ. Control*, 9, 213, 1980.

39. Chan, K. Y., Davey, B. B., and Geering, H. R., Interaction of treated sanitary landfill leachate with soil, *J. Environ. Qual.*, 7, 306, 1978.

40. Griffen, R. A. and Shimp, N. F., Attenuation of Pollutants in Landfill Leachate by Clay Minerals, Illinois State Geological Survey, Environmental Geological Notes, Number 78, U.S. EPA-600/2-78-157, Washington, DC, 1978.

41. Campbell, D. J. V., Parker, A., Rees, J. F., and Ross, C. A. M., Attenuation of potential pollutants in landfill leachate by Lower Greensand, *Waste Manage. Res.*, 1, 31, 1983.

42. Harmsen, J., Identification of organic compounds in leachate from a waste tip, *Water Resour.*, 16, 699, 1983.

43. Reinhard, M., Goodman, N. L., and Barker, J. F., Occurrence and distribution of organic chemicals in two landfill leachate plumes, *Environ. Sci. Technol.*, 8, 953, 1984.

44. Sawhney, B. L. and Kozloski, R. P., Organic pollutants in leachates from landfill sites, *J. Environ. Qual.*, 13, 349, 1984.
45. Venkatarami, E. S., Ahlert, R. C., and Corbo, P., Biological treatment of landfill leachates, *Crit. Rev. Environ. Control*, 14, 333, 1984.
46. Wilson, J. T., Enfield, C. G., Dunlop, W. J., and Cosby, R. L., Transport and fate of selected organic pollutants in a sandy soil, *J. Environ. Qual.*, 10, 501, 1981.
47. Harmsen, J., Possibilities and limitations of landfarming for cleaning contaminated soils, in *On Site Bioreclamation*, Hinchee, E. and Olfenbuttel, R. F., Eds., Butterworth-Heinemann, Stoneham, MA, 1991, 255.
48. Lui, K. H., Enfield, C. G., and Mravik, S. C., Evaluation of sorption models in the simulation of naphthalene transport through saturated soils, *Groundwater*, 29, 685, 1991.
49. Green, W. J., Lee, G. L., and Jones, R. A., Clay-soils permeability and hazardous waste storage, *J. Water Pollut. Control Fed.*, 53, 1347, 1981.
50. Alexander, M., Biodegradation of chemicals of environmental concern, *Science*, 211, 132, 1981.
51. Morgan, P. and Watkinson, R. J., Microbial methods for the clean-up of soil and ground water contaminated with halogenated organic compounds, *FEMS Microbiol. Rev.*, 63, 277, 1989.
52. Scow, K. M., Schmidt, S. K., and Alexander, M., Kinetics of biodegradation of mixtures of substrates in soil, *Soil Biol. Biochem.*, 21, 703, 1989.
53. Scow, K. M. and Hutson, J., Effect of diffusion and sorption on the kinetics of biodegradation: theoretical considerations, *Soil Sci. Soc. Am. J.*, 56, 119, 1992.
54. van Loosdrecht, M. C. M. and Zehnder, A. J. B., Energetics of bacterial adhesion, *Experientia*, 46, 817, 1990.
55. van Loosdrecht, M. C. M., Lyklema, J., Norde, W., Schraa, G., and Zehnder, A. J. B., Electrophoretic mobility and hydrophobicity as a measure to predict the initial step of bacterial adhesion, *Appl. Environ. Microbiol.*, 53, 1898, 1987.
56. van Loosdrecht, M. C. M., Lyklema, J., Norde, W., Schraa, G., and Zehnder, A. J. B., The role of bacterial cell wall hydrophobicity in adhesion, *Appl. Environ. Microbiol.*, 53, 1893, 1987.
57. Marshall, K. C. and Cruickshank, R. H., Cell surface hydrophobicity and the orientation of certain bacteria at interfaces, *Arch. Microbiol.*, 91, 29, 1973.
58. Fletcher, M., The attachment of bacteria to surfaces in aquatic environments, in *Adhesion of Microorganisms to Surfaces*, Ellwood, D. C., Melling, J., and Rutter, P., Eds., Special Publication of the Society for General Microbiology, Academic Press, New York, 1979, 87.
59. Rosenberg, M. and Kjelleberg, S., Hydrophobic interactions: role in bacterial adhesion, *Adv. Microb. Ecol.*, 9, 353, 1986.
60. Marshall, K. C., Adhesion and growth of bacteria at surfaces in oligotrophic habitats, *Can. J. Microbiol.*, 34, 503, 1988.
61. Marshall, K. C., Mechanisms of bacterial adhesion at solid-water interfaces, in *Bacterial Adhesion, Mechanisms and Physiological Significance*, Savage, D. C. and Fletcher, M., Eds., Plenum Press, New York, 1985, 133.

62. Mozes, N., Marchal, F., Hermesse, M. P., Van Haecht, J. L., Reuliaux, L., Leonard, A. J., and Rouxhet, P. G., Immobilization of microorganisms by adhesion: interplay of electrostatic and nonelectrostatic interactions, *Biotechnol. Bioeng.,* 30, 439, 1987.

63. Lane, W. F. and Loehr, R. C., Estimating the equilibrium aqueous concentrations of polynuclear aromatic hydrocarbons in complex mixtures, *Environ. Sci. Technol.,* 26, 983, 1992.

64. Weissenfels, W. D., Klewer, H.-J., and Langhoff, J., Adsorption of polycyclic aromatic hydrocarbons (PAHs) by soil particles: influence on biodegradability and biotoxicity, *Appl. Microbiol. Biotechnol.,* 36, 689, 1992.

65. Barriault, D. and Sylvestre, M., Factors affecting PCP degradation by an implanted bacterial strain in soil microcosms, *Can. J. Microbiol.,* 39, 594, 1993.

66. Characklis, W. G. and Marshall, K. C., *Biofilms,* John Wiley & Sons, New York, 1990.

67. Cunningham, A. B., Bouwer, E. J., and Characklis, W. G., Biofilms in porous media, in *Biofilms,* Characklis, W. G. and Marshall, K. C., Eds., John Wiley & Sons, New York, 1990, 697.

68. Slater, J. H., Mixed cultures and microbial communities, in *Mixed Culture Fermentations,* Bushell, M. E. and Slater, J. H., Eds., Academic Press, London, 1981, 1.

69. Atlas, R. M. and Bartha, R., *Microbial Ecology,* 2nd ed., Benjamin/Cummings, Menlo Park, CA, 1987.

70. Mortland, M. M., Mechanisms of adsorption of nonhumic organic species by clays, in *Interactions of Soil Minerals with Natural Organics and Microbes,* Huang, P. M. and Schnitzer, M., Eds., Soil Science Society of America Special Publication No. 17, Madison, WI, 1986, 59.

71. Santoro, T. and Stotzky, G., Sorption between microorganisms and clay minerals as determined by the electrical sensing zone particle analyzer, *Can. J. Microbiol.,* 14, 299, 1968.

72. Gordon, A. S. and Millero, F. J., Electrolyte effects on attachment of an estuarine bacterium, *Appl. Environ. Microbiol.,* 47, 495, 1984.

73. Stotzky, G., Influence of soil mineral colloids on metabolic processes, growth, adhesion, and ecology of microbes and viruses, in *Interactions of Soil Minerals with Natural Organics and Microbes,* Huang, P. M. and Schnitzer, M., Eds., Soil Science Society of America Special Publication No. 17, Madison, WI, 1986, 305.

74. Fontes, D. E., Mills, A. L., Hornberger, G. M., and Herman, J. S., Physical and chemical factors influencing transport of microorganisms though porous medium, *Appl. Environ. Microbiol.,* 57, 2473, 1991.

75. Harter, R. D. and Stotzky, G., Formation of clay-protein complexes, *Soil Sci. Soc. Am. Proc.,* 35, 383, 1971.

76. Harter, R. D. and Stotzky, G., X-ray diffraction, electron microscopy, electrophoretic mobility and pH of some stable smectite-protein complexes, *Soil Sci. Soc. Am. Proc.,* 37, 116, 1973.

77. Collins, Y. E. and Stotzky, G., Heavy metals alter the electrokinetic properties of bacteria, yeasts, and clay minerals, *Appl. Environ. Microbiol.*, 58, 1592, 1992.
78. Douben, P. E. T. and Harmsen, J., Diffuse verontreiniging van het landelike gebied—Sanering: doen of niet doen?, *Bodem,* 4, 157, 1991.

CHAPTER 7

Revegetation of Landfill Sites

Peter J. K. Zacharias

CONTENTS

I. INTRODUCTION

Revegetation of waste sites is a common requirement in legislation governing the disposal of anthropogenic by-products. In many countries the treatment of industrial waste and mine spoils requires "rehabilitation" of the site to its preuse standard (restoration). In the case of mine and industrial wastes this usually involves some form of dumping, landscaping, and revegetation using appropriate agricultural practices. This scenario is usual in situations where no important phytotoxic elements are present. Descriptions of the design of dumps as well as procedures for rehabilitating the site are readily available.[1-3] Suggestions on the management and maintenance of the site are also provided in some texts.[1] In most cases, once the dump has been successfully covered, revegetation follows either naturally or via agricultural-type inputs (machinery, seeding, fertilizer, etc.).[1] This is possible because the dump (or waste material) is essentially inert (with respect to plant growth) as it does not contain or produce toxic products.

0-87371-968-9/95/$0.00+$.50
© 1995 by CRC Press, Inc.

In the case of landfills, however, the problem is entirely different. A scenario for the reclamation of the site must include the objective of containing the "material" within the landfill. This is because the processes that take place after the compaction and covering of the waste give the site entirely new characteristics. Many of the products of these processes are toxic to several life forms, including plants. Most authors who work in the area of landfills are preoccupied with means of containing these products and many suggest engineering-type solutions.[4-7] The advent of geomembranes and other synthetics used in regular civil engineering practices has largely made these containment designs reasonably successful,[3,5,6] although questions are now being raised as to the long-term efficacy of these measures.

Unfortunately, many landfills have a tendency to subside, as a result of consolidation of material and microbial activity within the refuse,[6,8,9] and this may lead to a puncturing of the containment membrane. Senior et al.[11] recently argued that the challenges to plants from the products of the refuse fermentation (gas and leachate) may render reclamation programs merely cosmetic. If their argument is correct (there is currently little evidence to dispute it), it is clear that the technology for revegetation so widely practiced for the mining industry[1-3] will not be transferable to most landfill situations.

Apart from the physical and chemical limitations to plant growth, many mine dump reclamation projects have "super" budgets. Some figures for reclamation programs in South Africa approach 30,000 to $40,000 per hectare. These financial inputs are enormous relative to the productive potential of the land. Unfortunately, success often depends on huge investment in the "right" technology. In the case of municipal landfills, budgets for rehabilitation are unlikely to approach these levels. Consequently, low-cost options need to be found.

From the above argument it appears that specific technology relating to the revegetation of landfills needs to be developed. The purpose of this chapter, therefore, is to consider some options. An electronic search of nearly 13 years of abstracts yielded very few studies concerning revegetation of landfills per se. This indicates a lack of fundamental science in this area so the majority of what follows will be speculative. In a recent study of 600 closed sanitary landfills in Finland, Saarela[10] concluded that no directives were available for a Code of Practice for closing sites. As many European and American landfills are in excess of 50 years old[10] the development of knowledge to underpin such guidelines should be a priority.

The focus of this chapter relates directly to landfills associated with an urban environment and, thus, exclude specialized waste dumps[3] for highly toxic substances or nuclear material. A characteristic of such dumps is that they are usually managed by local authorities with "tax dollar" budgets and confined within city limits. This has important consequences for the overall planning as it affects the availability of materials for revegetation programs. Usually, the manager responsible for the site will have an engineering background and will operate from a philosophy of "engineering solutions". This policy makes sense for organizations with limited budgets because the major inputs during the life of the landfill are engineering (i.e., design, earth moving, dumping, spreading, recycling, landscaping, and so forth). As in the field of mining, few local authorities will employ a specialist biologist to plan and direct rehabilitation programs even if such expertise was available. It may be for this reason that so few studies have been reported in the literature in this specialist area.

Apart from management of the site itself, the plants too will experience problems.[11-13] These relate mainly to toxic leachate and gas. The leachate is often nutrient rich and, theoretically, is ideal for plant growth, but because of associated organic acids, is usually phytotoxic. Landfill gases such as ethylene and hydrogen sulfide are toxic to plants and it appears that considerable research effort will be required to overcome this problem.[11]

In any rehabilitation program involving vegetation, the most important component is the soil used for the final capping of the fill (Chapter 6). In an urban situation this is often in short supply or, if available, may be too expensive to buy and transport to the site. As a result, material that is used to cover the site prior to revegetation is often of poor or variable quality. Much of this material may have been dumped and have as its source construction sites (e.g., builders' rubble). The problems associated with revegetation, therefore, are exacerbated by the lack of suitable growth media. It seems inevitable that an integrated approach, planned by a range of biological specialists, will be required to develop technologies to provide suitable species for revegetation of landfill sites.

II. REVEGETATION REQUIREMENTS

The goal for the reconstruction of a suitable medium for revegetation is to provide a capping that is deep and as favorable to root growth as is necessary to achieve desired plant performance.[2] Several factors are important, including plant production, basal area of plants, protection from erosion, appropriate compaction, and others. In some cases the removal of water by plants is also important as this will reduce percolation and excessive seepage.

This list of requirements suggests that the operator needs to reconstruct deep soils to allow unchallenged root growth. In a mining context this may be possible but because of the characteristics of landfill is unlikely to occur in most cases.[10]

A. HANDLING GROWTH MEDIA

It is important at the design stage of a landfill development that this is carried through to the closure phase of the project. The standard of the design and its scope will significantly affect the quality of the reclamation program.[14]

It makes sense, therefore, that any suitable capping material is removed from the site prior to refuse emplacement. It is particularly important that the topsoil that is removed is carefully stored and protected.[2] To ensure that the soil remains suitable for plant growth it must be protected, while stock piled, against contamination and erosion.

B. REVEGETATION

In theoretical terms, once the landfill has been capped and any drainage and gas harvesting structures are in place, the area could be left to undergo a process of secondary succession. This is the classical view of land restoration rather than reclamation, and is based on so-called ecological principles.[15] The model is low cost and attractive. However, it presupposes that the site is free of leachate and gas problems[11-13] which are common features of domestic and industrial landfills. In many cases, seed banks are nonexistent or have been destroyed during soil storage prior to use as closing material. Revegetation programs will in all probability require inputs of seeds and fertilizer at least in the initial phases.[2] Once suitable vegetation has been established and plant propagules are moved naturally onto the site, the fertilizer inputs can stop.

It appears, therefore, that a specific program will need to be developed for each site as conditions for growth are likely to be unique at each site.[16]

III. THE RESEARCH CHALLENGE

If we accept that features (e.g., deep capping and few toxic elements) common on high-cost mine dump rehabilitation projects will not occur in most landfills, we are faced with the challenge of developing suitable technologies. A review of papers delivered at the recent (1993) International Landfill Symposium[17] shows that considerable effort is being made to harness products from landfills and to exploit recycling technology.[11] This is obviously an environmentally responsible approach and will become increasingly more important as legislation, currently applicable to the mining industry,[2,14] is extended to include all reclamation programs. An inevitable consequence of more embracing legislation will be the need to pay specific attention to the revegetation of landfill sites once they are closed. As many are in operation for several years, the revegetation program will last the life of fill and for many years beyond.

The main responsibility of the biologists then is to determine which characteristics that affect plant growth are common to "most" landfills. Once these have been established, research efforts should concentrate on finding plant species which are tolerant of the conditions. In addition to this, projects similar to those described by Senior et al.[11] involving mycorrhizae should be further developed.

An important aspect of the revegetation process is choice of suitable species. Those authors who have addressed revegetation aspects of landfills have tended to favor tree species. Given that most modern cover designs include a "biotic barrier"[6] or "geomembrane"[4] the deep penetrating roots of tree species are not desirable. Grass species, because of their shallow and fibrous root systems, are likely to be ideal reclamation species for landfill sites. Grasses have several advantages over tree species. Their fibrous roots contribute rapidly to organic matter build-up and this is an important component of nutrient cycling and storage.[2] The lack of deep penetration of grass root systems means that the integrity of barriers designed to contain landfill products is not at risk. Erosion by wind and water is most effectively controlled by grass plants because they tend to form a closed canopy and relatively high basal cover. Should the area stabilize, grasses form the ideal "pioneer" community as a precursor to the development of a more complex vegetation structure. As the complexity of the vegetation increases, the site may be released to other forms of land use, e.g., recreation. This, however, will only be possible in situations where the site can be declared safe for humans.

IV. CONCLUSION

It is clear from a search of the literature and the discussion above that landfill sites have unique properties which are governed by the interactions between the first-tier variables which are specific to each site. Because of this, any general Code of Practice for reclamation, similar to that in the mining industry, is unlikely to be achieved. As in all ecological, rather than engineering, problems *uniqueness is ubiquitous.*

The challenge for the researchers is to find what minimum levels of understanding are required at each site. On the basis of this information a coordinated interdisciplinary team will be needed to achieve the following:

1. Determine additional information peculiar to that site.
2. Select suitable capping material and appropriate amelioration (liming, fertilizer).
3. Select species tolerant of any specific phytotoxic challenge.

4. Design and landscape the "final" landfill characters (e.g., care of water management, leachate, and gas).
5. Develop management plans for aftercare once the site is closed.

It should also be obvious that such a procedure will require expensive specialists at each survey. This, however, will not be sustainable in the long term due to costs and limited manpower and some form of artificial intelligence will be required. Recent developments in decision support systems[18] must be employed to assist the landfill manager with the rehabilitation program. Of course, no suggestion is made that a computer will replace the specialist team, although computer models can be used to assist the manager with general practice and decision making. Project leaders responsible for landfill research will need to not only broaden their teams to include expertise across a wide spectrum of engineering and biological disciplines but also work with decision support specialists.

REFERENCES

1. Williamson, N. A., Johnson, M. S., and Bradshaw, A. D., *Mine Wastes Reclamation: The Establishment of Vegetation on Metal Mine Wastes,* Mining Journal Books, London, 1982.
2. Vogel, W. G., *A Manual for Training Reclamation Inspectors in the Fundamentals of Soils and Revegetation,* USDA, Berea, KY, 1987.
3. Salomons, W. and Förstmer, U., Eds., *Environmental Management of Solid Waste: Dredged Material and Mine Tailings,* Springer-Verlag, New York, 1988.
4. Rollin, A. L., Lafleur, J., Lemieux, R., Wyglinski, R., and Zanesau, A., Capping of mining wastes using textured HDPE geomembranes, in *Proc. Sardinia 93 Fourth Int. Landfill Symp.,* Cagliari, Italy, 1993, 243.
5. Brunner, D., Dirgo, J., and Gandhi, B., Design and construction of a landfill cover in a flood plain, in *Proc. Sardinia 93 Fourth Int. Landfill Symp.,* Cagliari, Italy, 1993, 1636.
6. Wallace, R. B., The stability and integrity of landfill closure systems, in *Proc. Sardinia 93 Fourth Int. Landfill Symp.,* 1993, 253.
7. Anon., *Reclamation 83,* Industrial Seminars, Kent, England, 1983.
8. Dodt, M. E., Sweatman, W. B., and Bergstrom, W. R., Field measurement of landfill settlements, in *Geotechnical Practice for Waste Disposal,* ASCE, New York, SP13, 406, 1987.
9. Landva, A. O. and Clark, J. I., Geotechnics of waste fills, in *Geotechnics of Waste Fills: Theory and Practice,* ASTM STP 1070, American Society for Testing and Materials, Philadelphia, PA, 1990, 86.
10. Saarela, J., Types and costs of covers of closed landfill sites, in *Proc. Sardinia 93 Fourth Int. Landfill Symp.,* Cagliari, Italy, 1993, 235.
11. Senior, E., Watson-Craik, I. A., and du Plessis, C. A., Landfill site restoration, *Resource,* October, 23, 1992.

12. Williams, N. M., Senior, E., and Zacharias, P. J. K., Determination of minimum toxicity thresholds of landfill leachate molecules to selected site reclamation species, in Abstracts South African Society for Microbiology (Natal) VI Ann. Symp., Durban, South Africa, 1993, 35.

13. Grantham, G. and Young, P., Practical experience in landfill remediation, in *Proc. Sardinia 93 Fourth Int. Landfill Symp.,* Cagliari, Italy, 1993, 1533.

14. Wilson, G., Integrate landfill restoration, in *Proc. Sardinia 93 Fourth Int. Landfill Symp.,* Cagliari, Italy, 1993, 1549.

15. Shaw, P. J., Development of vegetation in landfill environments, in Reclamation 83, Proc. Int. Land Reclamation Conf., Grays, England, 1983, 503.

16. Klatkin, E. M. and Watkin, J., Reclamation programmes for mining and industrial wastes, in Reclamation 83, Proc. Int. Land Reclamation Conf., Grays, England, 1983, 483.

17. Christensen, T. H., Cossu, R., and Stegmann, R., *Sardinia 93: Barrier Systems, Environmental Aspects, Upgrading and Remediation, Siting, Monofills, Effects of Waste Pretreatment, Landfilling in Developing Countries,* CISA Environmental Sanitary Engineering Centre, Cagliari, Italy, 1993.

18. Fedra, K., Integrated information systems for water resource management, in Watercomp '93: Computing for the Water Industry Today and Tomorrow, Melbourne, Australia, 1993, 277.

Index